T0271207

# Resilient Space Systems Design: An Introduction

# Resilient Space Systems Design: An Introduction

### Authored by

RON BURCH

**CRC Press**
Taylor & Francis Group
Boca Raton London New York

CRC Press is an imprint of the
Taylor & Francis Group, an **informa** business

CRC Press
Taylor & Francis Group
6000 Broken Sound Parkway NW, Suite 300
Boca Raton, FL 33487-2742

Copyright © 2020 by Taylor & Francis Group, LLC
CRC Press is an imprint of Taylor & Francis Group, an Informa business

No claim to original U.S. government works

Printed on acid-free paper

International Standard Book Number-13 978-0-367-14848-5 (Hardback)

**Visit the Taylor & Francis website at**
http://www.taylorandfrancis.com

**and the CRC Press website at**
http://www.crcpress.com

# Contents

# Preface

The somewhat arcane topic of space resilience has actually been with us for several decades now in one form or another. During that time, it has gone by different names and has been addressed in a number of different ways. Commercial and civil space designers have had to consider the impact of hazards to satellites and planetary probes that include solar flares and space weather, the natural radiation environment, and other threats. Designers of military satellites have had to consider nuclear effects and intentional interference in addition to the natural hazards. Now, as space grows more congested, additional threats are posed by neighboring satellites and the growing problem of space debris. The study of resilience encompasses all of these concerns as well as the mitigation of emerging threats to both satellites and the associated ground system, such as cyber attacks.

The origins of this book go back several years when the U.S. government began taking these emerging and escalating threats to their space infrastructure much more seriously. The desire for more "resilient" space assets began to be discussed both privately and publicly. As I began to consider the implications of designing future space systems to survive and operate in future threat environments, it became clear that the definition of resilience was unclear to many, and that there was little consensus in the space community. Further, in trying to design for resilience, there were few accepted metrics and no obvious or accepted methods to calculate this parameter. The Department of Defense began to publish documents that attempted to clarify these definitions beginning around 2010. From that foundation, the contents of this book sprang.

Those who read this book hoping to find a narrative piercing the veil of secrecy and providing visibility into the detailed efforts of the United States and its allies to provide resilience to future systems will be disappointed. Rather, my intent is to provide those interested in designing future space systems with an introduction to the definitions, approach, tools, and methods useful in making these systems more resilient. All of the information and data taken from the real world is provided as concrete examples, the vast majority of which is derived from publicly available information, with sources provided as references.

Furthermore, even the specific methods here are not presented as the one true approach to designing resilient systems. Indeed, my references include alternate approaches and the reader is encouraged to continue their education to determine which approach is most appropriate for their application. However, this book is founded on fairly simple but immutable facts, with the necessary mathematics and substantiation, that will apply to all of these different approaches.

Finally, resilience is not a pure and mathematically precise discipline. This is because the accuracy of the evaluation of resilience is dependent upon the accuracy of the threat definition, which may be to some extent inferred from observable events. Without a precise knowledge of the threat, there will always be a level of uncertainty in the effectiveness of the threat as well as the threat mitigation, and thus the estimated resilience of the system. I do hope that the basic concepts presented in this book provide a certain level of guidance to those considering what it means to design for resilience and the trades that are best considered to accomplish it.

An Excel workbook is provided as a supplementary aid to enhance understanding of the mathematics and equations presented in the book. The worksheets implement many of these equations and enable the user to easily manipulate them to gain insight into their use and characteristics. The equations themselves are embedded into each worksheet with accompanying references to the relevant sections in the book.

My sincere thanks to those who have contributed their ideas and perspectives and reviewed the content provided here. My thanks also to those who have provided thought-provoking questions that led to deeper insights and better ways to describe the process of designing for resilience. I hope the result is an easy-to-follow guide to the topic of designing resilient space systems.

# About the Author

**Ron Burch** is the Director of Advanced Military Satellite Communications (MILSATCOM) for the Boeing Company's Space & Launch division. He has over 35 years of satellite systems design and development experience at the Boeing Company and Hughes Aircraft Company. He is an acknowledged subject matter expert in the emerging discipline of space system resilience and has spoken internationally on the subject. His roles have included RF and digital subsystem and payload design, systems engineering, space technology development, and leadership positions including program management. Ron received a Bachelor's degree in electrical engineering (BSEE) from California State University, Fresno, and a Master's degree (MSEE) from the California Institute of Technology (Caltech) with an emphasis in communications science. He has published numerous technical papers and is named on two U.S. patents.

# Introduction

**Resilient**
adjective *re·sil·ient* \ri-'zil-yənt \
Tending to recover from or adjust easily to misfortune or change

— *Merriam-Webster Dictionary*

The word "resilient" has come into common use with increasing regularity. Athletes and their sports teams, countries and their economies, and even microbes have all been described as resilient. In most cases, the use is more qualitative than quantitative: Resilience is expressed as more of a property or an attribute than a specific number or value. While this is acceptable for everyday use, the lack of precision poses a challenge for engineers tasked with designing products that are increasingly required to be "resilient." Engineers need clear definitions, mathematical models, and the ability to specify and measure attributes to ensure that the end product or service performs as advertised and actually *is* resilient to threats when operating over a range of environments and scenarios.

The purpose of this book is to provide a comprehensive introduction to the basic principles of resilient space system design, including foundational concepts and methods. The early chapters of the book provide introductory information and definitions upon which to build, including a review of the components and characteristics of a space system, the definition of resilience as applied to space, and an overview of threats and threat mitigation. A detailed discussion of design methodology begins in Chapter 5, starting with the derivation of an equation for calculating resilience based on four key resilience attributes: avoidance, robustness, recovery, and reconstitution. This capstone equation, which is straightforward yet highly useful, is examined in detail, including the modeling of each of the key resilience coefficients. The remaining chapters cover design trades and apply the methodology to illustrate the resilient design process through the use of examples. The space system examples provided in this book are largely simplified satellite systems, chosen to best illustrate the design for resilience approach and methodology. The content is applicable to other types of space systems and the modeling and mathematics presented are general and extensible.

In contrast to much of the existing literature, this book focuses on principles of the *design* of resilient space systems rather than simply analysis. There is a distinction between performing a system resilience analysis and performing a system design. Many of the calculations are the same, but the intent and processes are different. In the first case, an existing system is evaluated for resilience by analyzing its predicted behavior in the presence of one or more threats. However, a system design activity begins with the system performance, cost, and threat requirements that form the starting point for system designers to minimize the impact of the threats and maximize the resilience of the system. The threats can drive the system design as much as other performance specifications, requiring the addition of threat mitigation features to improve resilience prior to building the system. It is thus desirable that the approach to calculating resilience provide a tool for system designers, with parameters tied to implementation details under their control, rather than an equation that does not assist in design trades.

The design process, while disciplined, retains a certain amount of creativity on the part of the designer and this book does not mean to imply that the contents fulfill all of the needs of someone wishing to learn resilient space system design. Instead, the goal is to provide concepts, tools, and approaches useful in incorporating resilience as a part of the established system design process. This is not a book about redefining the discipline of systems engineering, but rather merely to augment it.

The recent heightened interest in space system resilience is rooted in the recognition and acceptance of the fact that these systems, once considered virtually untouchable, are being recognized as much more vulnerable than was previously believed [1]. For most of the Space Age, the space domain in particular has been considered a sanctuary for satellites, safe from terrestrial threats, and subject only to certain very rare hazards such as orbital debris and extreme solar weather. As a result, the primary concern of the designers of these systems has been designing to well defined natural environments and maximizing the reliability of the satellites themselves and the associated gateway and command and control ground stations.

Recent events in the twenty-first century have rendered this view obsolete, resulting in the need for new approaches to designing, deploying, and operating future space systems. In addition to being highly reliable, these systems must now also be resilient when confronted with existing and emerging threats. As space itself has become more "congested, contested, and competitive," the range of credible threats, both to satellites and the associated ground segment, continues to increase.

At present there is still some ambiguity and disagreement as to the exact definition of resilience as it is applied to space systems. More frequently other words are invoked when describing the *qualities* of resilient systems: *flexibility, agility, scalability, modularity,* and *extensibility* are but a few of these, each providing some measure of resilience to a system. And as with resilience, these words are more descriptive than prescriptive, lacking measurable values. Measuring a system's flexibility is a challenge, for example.

This book takes a more formal, stepwise approach to resilient space system design, providing clear definitions and building out a methodology and framework using precise terms and mathematics. While there may not be a consensus within the community, there are clearly many basic underlying principles that are common to many published approaches to the subject.

As a result, the details of the specific methodology are less important than the foundational concepts. This book begins with first principles, establishes some fundamental definitions, and presents a basic methodology and an associated design process for imparting resilience to space systems. This is by no means intended to be the final word on the subject, simply a guide to approaching the design of resilient space systems.

Many modern space systems are, in fact, *information* systems. They route data from sources to destinations (users). Some systems generate this data internally while others receive it from external sources for transmission to users or consumers of that data. Space system capabilities include communications, imaging, navigation, and sensing. A single system can provide one or more of these simultaneously. The examples in this book will be restricted to more conventional near-Earth applications, but the methodology may be extended to any space mission regardless of its capability. A resilient space system is capable of continuing to deliver its data or service continuously, even when impacted by partial system failure, adverse conditions, or hostile actions. The U.S. government also refers to this as continuity of operations (COOP) [2]. It is this concept of a resilient system being able to limit the depth and duration of service degradation or interruption as well as the means by which this is accomplished that is key. In particular, time is a critical factor and minimizing outages is a prime concern.

While resilience might not yet be a universally accepted systems engineering discipline, this is not to say that engineers have not been designing for resilience, but rather that a comprehensive approach has not yet been widely adopted. Instead, engineers have addressed specific and selective threats and conditions in their design process. Terrestrial examples include designing civil structures to sustain extreme weather and natural disasters, such as hurricane force winds, earthquakes, or tsunamis. In space, engineers design satellites to operate in harsh radiation and thermal environments. U.S. strategic satellites, born in the 1970s and 1980s, have long been designed to be "survivable" in the presence of nuclear events in space. For these systems, clear threats were identified, requirements and specifications for nuclear survivability were defined, design guidelines were developed, and hardened satellites were built and tested and continue to be operated today. So, designing for resilience is nothing new, but the range of credible threats has expanded.

Today's challenge is to apply a more general approach for designing for resilience while considering a wider range of threats and mitigation techniques to space systems. In the past, the system designer has been largely concerned with performing the key trade between system cost and performance. However, now resilience has joined these two trade parameters as part of a three-way trade (Figure 0.1). It is no longer enough to design and deploy a high-performance space system; those capabilities must also be maintained in the presence of one or more threats over some period of time. To date this mission duration for a typical space system has been in excess of

**FIGURE 0.1**
The new space system trade space includes resilience.

a decade, and with an increased rate of threat escalation, maintaining resilience over such a long period of time is becoming ever more challenging.

The relative values for cost, performance, and resilience depend upon the type of system and its service or mission, as shown in Figure 0.1. Some systems are experimental, with more emphasis on the demonstration of a capability over a shorter period of time with little residual operations required; these systems often prioritize affordability. Mission-critical systems provide a capability that might not always offer the highest performance, but prioritize assured access and mission assurance, and thus resilience. Finally, large, sophisticated satellite systems (sometimes called *exquisite systems*) exist primarily to deliver unique capabilities and focus more on performance. These most important requirements (MIRs) must be established when entering the design process for any new space system to set priorities for the system designers.

There are many different ways to impart resilience to a system, and the cost and performance varies depending on which solution is chosen. Increasing system resilience usually results in an incremental cost to the system (or "resilience tax") and/or some performance degradation as well. The designer's task is to minimize these costs while maximizing the system resilience to the specified threats.

While it is tempting to conclude that only military or government space systems are the targets of hostile actions, this is not necessarily true. Although government-owned systems that provide nations with military advantages are certainly more likely to be targets of hostile actions, many nations rely on commercial satellite services for a variety of purposes, including civil and military uses, thus making them potential targets as well. Political motives can also result in the targeting of space systems. As more countries avail themselves of satellite communications (SATCOM) and overhead imagery for both economic and military uses, adversaries have greater motivation to disrupt these systems for their own gain. Today the technological and cost barriers to entry for acquiring and deploying some of these threats have been lowered.

And hostile actions are not the only concern. Adverse conditions are also increasing due to greater congestion in space and the increasing threat posed by orbital debris which can endanger both government and commercial space systems. Conditions such as space weather (e.g., solar flares coronal discharges), and terrestrial weather, which can interrupt space-to-ground radio frequency (RF) communications links, can also impact system performance. RF interference, a common threat, can be either intentional or unintentional. As more space systems become operational, the congested environment greatly increases the likelihood of unintentional interference in space. As a result, planning simply for hostile actions is not enough to assure resilience in the future.

In searching for a blueprint for an approach to resilience, one well-established engineering discipline provides a good starting point: reliability engineering. Reliability engineering has become an established discipline within space systems engineering, one which is built on clear and specific definitions, engineering tools, and mathematical models. As with resilience, even the general public has an intuitive concept of what a reliable product or service is: something free from failure due in part to good design and manufacturing, resulting in freedom from defects. Simply put, reliability engineering exhibits much of what is desired in describing a similar framework to address resilience.

The key difference between reliability and resilience is that reliability is defined by the system's response to failures *internal* to the system while resilience is defined by the system's response to *external* threats or conditions. In both cases, the design engineer is concerned with providing uninterrupted capability or service; the difference is the source and nature of the potential outage as well as the available mitigations.

Reliability engineering provides designers with the mathematical models to make quantitative predictions of the system reliability based on component failure rate data and the system's design and operation. These reliability estimates are derived from failure rate models based on measured and historical component test and operational data which then become a part of the system's operational availability. The reliability metric is a probability of system (or mission) success (or failure) over some defined time period of interest such as satellite design life. In Chapter 5, a methodology is provided using system design information coupled with a description of external threats to yield an expected value of the residual system capability that can be calculated for a specific threat or threat scenario, similar to a reliability prediction.

The true system response to a threat is unknown until and unless the threat is realized and the actual system response to the resulting event is observed. The system resilience prediction is, with the exception of certain special cases, often probabilistic, including a number of factors that influence the resulting residual system capability. Though the input data is different, for both reliability and resilience the goal is to quantify the impacts to the system's capability level using analytical techniques to obtain a best estimate under the specified operational conditions.

Reliability is also a parameter that can be allocated among the many system components, which is a desirable attribute for system design. One of the key system design tasks is the allocation of key performance parameters to system elements. The resilience metric presented in this book is also a quantity that can be allocated or assigned to the various parts of the system. Although the allocation process is not identical, the concept is similar. As with reliability, the resilience of each system element to each specific threat can be determined to provide the required system-level resilience to the threat. Through this process the threat mitigation features necessary to increase the resilience of the system can be identified and their location be defined.

As with any design and analysis activity, the fidelity of the results is dependent upon the quality and quantity of the input data. Many inputs, such as threat assessments, are uncertain to some extent. It is important to keep in mind that, during the design process, candidate system architectures are usually compared in a relative sense, so that as long as the same criteria are used, absolute values may be less important. If desired, error bars can be placed on the coefficient values to understand the overall uncertainty and sensitivity to parameters in the final resilience value. The proposed equations are well behaved in that regard.

It quickly becomes apparent why accurately determining system reliability seems more achievable than resilience: System designers have extensive knowledge of how the system is designed, including the components that comprise it, how they are interconnected, their failure rates based on testing or history, and the environmental conditions over which the system is required to operate. Given sufficient data, a high-fidelity model can be created to predict the reliability of the entire system providing the required level of performance over a given period of operation. Designing for resilience, on the other hand, requires accurate knowledge of threats and conditions external to the system. This information can be much more difficult to obtain and in many cases is likely to be less accurate. So, while reliability predictions are estimates with error bars, any similar estimate of resilience is likely to have an even wider range of error due to the inherent uncertainty in the input data.

The defining aspect of resilient design is the need to mitigate the effects of credible threats that have the potential to cause degradation to a space system. Threats are the external stimuli that the system must be resilient *against*, and they define the resilience requirements that influence the design of the space system. The definition and categorization of threats is a key driver in the system design process.

System designers rely on the knowledge of the range of threats to develop a suite of mitigation strategies. A threat is successful if when exercised it causes a performance degradation sufficient to cause an interruption of the capability delivered by the system. Some of the available mitigation approaches, such as distribution and redundancy, add resilience against multiple types of threats. Other approaches are specific to a single threat. Thus, a "more resilient" system tailored to the spectrum of potential threats can be designed. The more that the threats are understood, the better the estimate of *how* resilient the system is, thus enabling trade-offs of different mitigation approaches to ensure they are resilient enough to meet mission or service needs and that their cost is justified.

There is a tendency to focus on specific mitigations for specific threats, but it can be somewhat shortsighted when compared with the prior steps of evaluating resilience at higher levels to ensure that local solutions do not sub-optimize or preclude other approaches that might be more cost effective.

This book is *not* intended to delve into the details of the design of specific threat mitigation features, such as how to make a satellite more maneuverable to defeat a kinetic threat or methods of radiation hardening to survive nuclear effects. The examples provided herein are primarily "effects based," in that the system-level effect of the threat is indicated and evaluated rather than the underlying detailed analysis. This is because the detailed designs of such features vary substantially depending upon the exact threat and the nature of the system being designed and thus become a discipline for study all on its own. The science of protecting a satellite against the many adverse effects caused by a nuclear event alone is wide and deep. Those design principles are not unique to resilience design and analysis, though they will be the means of obtaining the resilience values used in resilience equations.

Instead, the primary goal of this book is to introduce the basics of resilient space systems design including the tools necessary to perform the trades among performance, cost, and resilience, and in doing so to streamline a complex problem and help the designer efficiently move forward with the design. The core concepts are in many ways simple, but the interactions, interdependencies, and large number and variety of components can come together to form a complex problem. Reducing that complexity is central to developing an accurate and practical design approach. By carefully applying resilient design techniques in a disciplined fashion, this complexity can be managed to enable optimization of the design resulting in a more resilient, capable, and affordable space system.

# 1

## The Space System

A space system delivers a capability or service through a combination of space-based elements such as satellites and terrestrial ground stations that communicate with one another. Its architecture and constituent components determine how it delivers these capabilities. The characteristics of each component also dictate both its vulnerabilities to threats as well as the available methods for adding resilience features that mitigate those threats. This chapter briefly reviews the most common space architectures and components and their characteristics and provides a discussion of the associated design considerations as they pertain to system level resilience.

## 1.1 Space System Components

Space systems are designed to provide one or more capabilities to its users by combining space and ground elements. Figure 1.1 shows the key components of a common space system architecture. The space system is comprised of a *space segment*, a *ground segment*, a *control segment*, a *network segment*, and, for communications systems (satellite communications, or SATCOM), a *user segment*, each of which is discussed in greater detail in the text. Sometimes a *launch segment* is also included, since a launch vehicle is necessary to deploy the space segment. Sometimes all of the ground-based components, including the control and network segments, are designated as the ground segment. Not every system includes every component shown in the figure, but most space systems include the space, ground, and control segment elements. Again, for SATCOM systems, the users must also possess terminals by which they communicate with one another, so that segment is also essential for those systems.

One or more satellites make up the space segment, with each satellite orbiting the Earth (for now only Earth-based systems are discussed) and hosting one or more *payloads*. These payloads perform the key functions that deliver space-based capabilities and can include communications subsystems, sensors, and other functional components. In military parlance, the payload provides the mission capability. These payloads, which can include antennas, receivers, transmitters, and digital and analog processors, are housed on a spacecraft assembly called the *bus*, which provides the mounting structure and all other required support functions to enable payload operations, including DC power, propulsion, thermal and attitude control, and telemetry and commanding.

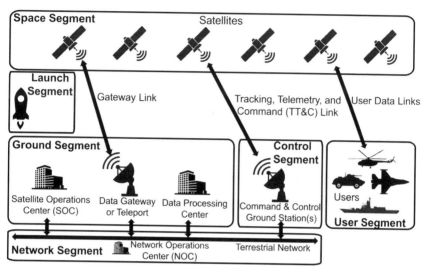

**FIGURE 1.1**
Key components of a space system architecture.

The satellite bus and the payload are usually commanded by the same means, although in certain cases the payload can be commanded by a separate command link from a different ground site than for the bus.

The ground segment consists of a satellite operations center (SOC) from which satellite operations are planned, managed, and executed. Sometimes a *mission planning segment* is depicted separately, but usually the SOC is the control point for the system, and particularly for the satellite constellation. The ground segment also includes terrestrial gateways that communicate directly with the satellites over RF communications links to send and receive data. Links that transmit data to the satellite are referred to as *uplinks*, while those used by the satellite to transmit data to the ground are called *downlinks*. These gateways must be in line-of-sight view of the satellites during these data transfers and are usually connected to the network segment for retransmission of the data to other locations. The gateway sites are often distinguished by large antennas, some of which must track moving satellites throughout the contact time. Some space systems, such as imaging systems, also include a separate data processing center to process received raw data prior to dissemination of processed imagery to users. These functions are usually considered a part of the ground segment, as are any additional ground-based data processing and data dissemination functions.

The control segment includes the satellite command and control sites that serve as access points for the SOC to implement commanding as well as reception of satellite health and status via telemetry downlinks. Sometimes this function is included in the ground segment, but often command and control sites, such as the U.S. Air Force Satellite Control Network (AFSCN), are shared by multiple satellite systems and are considered part of a separate system or network. Commercial satellite system operators use either owned or leased command and control sites in a similar manner, depending upon the architecture and requirements of their system.

The network segment includes a conventional terrestrial network that connects all ground components and can also be considered part of the ground segment, although more frequently this functionality is treated as a separate segment as parts of it are shared by multiple space systems as with the command and control sites. Network management is performed via the network operations center (NOC). This function can be housed in the same location as the SOC and maintains network connectivity. The NOC can also provide control of network-centric user terminals for SATCOM systems, which must log into the network prior to operating over the SATCOM system. Modern networks are generally Internet Protocol (IP)-based and sometimes use the Internet as part of the transport path using a virtual private network (VPN) to provide secure communications between network nodes.

SATCOM systems additionally include a *user segment* which consists of user terminals (radios) by which the communications service is accessed by the globally dispersed fixed and/or mobile users. If the terminal is operated while the user is moving, the terminal is referred to as an "on the move" (OTM) terminal. Terminals that are quickly set up following a move are referred to as "at the halt" (ATH) terminals.

Each system element is a potential threat target that, if realized, can result in the denial of part or all of the system capability to any or all of its users. Many systems provide multiple capabilities and a disruption to any of its elements can interfere with any or all of them. Designing a resilient space system requires that the designer identify these threats and include design features that mitigate their impact to maintain some minimum required capability at an acceptable level of performance.

It is important to understand how each element supports the end-to-end system capability and provides its own resilience to specific threats. Once this is known, the resulting loss of capability can be estimated for each element that is affected or impaired. However, the first step is defining the system's capabilities.

## 1.2 System Capability

Commercial and government space systems provide a wide variety of capabilities, from communications to Earth imaging to navigation. Operational systems (*and their primary capability*) include:

- Global Positioning System (GPS), Galileo; *Positioning, Navigation, and Timing*
- Wideband Global SATCOM (WGS); *Global Communications*

- Space-Based Infrared System (SBIRS); *Missile Warning*
- Geostationary Operational Environmental Satellite (GOES); Weather
- Space-Based Surveillance System (SBSS); *Space Situational Awareness*
- DigitalGlobe's WorldView satellites; *Optical Imaging*

Some of these systems are global, others are regional. Some move around the globe in low Earth orbits, while others dwell over a single spot on the Earth at a much higher altitude.

The *system capability* is the functionality or service provided by the system to its end users with some specified minimum level of performance. If the delivered performance is below this threshold, the user satisfaction is unacceptable. In government systems, this results in failure to support a mission. Resilience is thus a measure of the level of user satisfaction in the presence of one or more threats.

The system capability is best described by the performance or functional parameter that most exemplifies the service being provided. It is essential that this be the first step in designing for resilience so that the proper criteria are used. For a SATCOM system, the capability of interest might be the total capacity (or bandwidth) provided to users in a certain geographic area of the world. Or it could be the minimum data rate that is required to provide a certain level of service such as teletype, voice, or full-motion high-definition video to a specific user. Sometimes more than one capability is of interest and thus the resilience of the system for each capability must be considered. In this case, for the purposes of executing design trade studies, it is desirable to prioritize the capabilities of interest.

The system performance is often described in terms of an end-to-end quality of service (QoS) metric. Examples of this for modern digital communications waveforms are bit error rate (BER), channel error rate (CER), and packet error rate (PER). The acceptable values for these parameters are often specified as an instantaneous value or as a time-averaged value, depending upon the ability of the system to accommodate errors. The radar imagery community uses the Radar National Imagery Interpretation Rating System (RNIIRS) as a metric to describe synthetic aperture radar (SAR) image quality. Similar QoS metrics can be defined for other types of systems as well.

The capability metric should be measurable and a quantity that can be calculated such that any degradation due to the impact of a realized threat can be estimated based upon the properties of the threat, the system design, and the system concept of operations (CONOPS) which describes how the system is operated.

## 1.3 System Architectures

The manner in which a space system delivers one or more capabilities depends in large part upon the system architecture. The architecture is the structure created by the interconnection of system elements. System capabilities are often delivered through one of two primary architecture types: parallel or series, although in practice many systems are a hybrid combination of both.

### 1.3.1 Parallel Architectures

A common architecture is a parallel one in which an aggregate capability is provided by many elements simultaneously. The parallel architecture delivers its capability to users via multiple parallel and largely independent paths. In this case the total capability is the sum of that delivered by these elements, and the loss of any single element reduces the value of the total by some fractional amount but does not necessarily reduce it below the minimum required capability for a particular user of interest.

Figure 1.2 provides an illustration of a parallel architecture for a SATCOM system. In this example, some amount of capacity (bandwidth or throughput) is delivered to a geographic area via multiple satellites, each with equal capacity. In this simple example, four satellites are relaying signals to a group of users, with no need for ground control or routing. For this type of

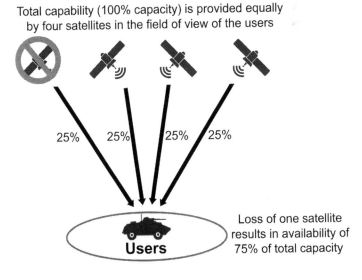

**FIGURE 1.2**
Example of a parallel (or aggregate) architecture.

architecture, the loss of one element results in a proportional loss of capability. If all four satellites deliver equal capacity, loss of one satellite results in a 25 percent loss of system capability.

To ensure adequate capability is maintained, a common design technique is to build margin into the system. In the case of the system in Figure 1.2, a fifth satellite could be included to mitigate the loss of a satellite. This would result in a system margin of 25 percent but might also add 25 percent to the cost of the space segment. More economical solutions might exist. Distribution strategies are discussed in depth in Chapter 6.

### 1.3.2 Series Architectures

The second common system architecture is represented by a chain of functional elements that must all be present and operating at certain minimum capability levels to ensure that the end-to-end capability is delivered. In contrast to the more distributed parallel architecture, if any one of these elements' capability level drops below a minimum value, then the end-to-end service is interrupted, and the system level capability becomes zero. As Figure 1.3 shows, this is a series architecture, where the failure of any single link in this chain can cause a degradation or complete system outage. If the Military Satellite Communications (MILSATCOM) Gateway is disabled as shown, no service or product is delivered to the system users.

Series architectures can be more fragile due to the requirement to maintain functionality of all of the elements in the chain, not allowing any element loss. If certain elements are judged to be unacceptably vulnerable, steps must be taken to increase those elements' resilience. Examples include added physical protection or the addition of redundancy at the element level.

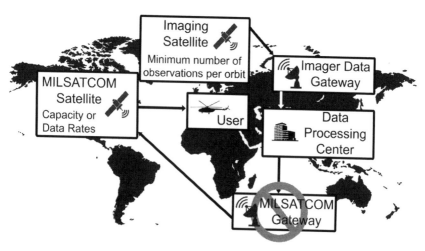

**FIGURE 1.3**
Example of a series architecture.

In this example, an imaging satellite system must provide some minimum performance by delivering imaging data of some minimum quality to one or more end users. The imaging satellite is the source of the data that is transmitted to a ground gateway that then forwards it to a data processing center via a terrestrial network. The center then transmits the post-processed imagery via the network to another ground station for uplink to a MILSATCOM satellite that relays it to the users. If the capability of any one of these system elements is disrupted, the end-to-end system capability is interrupted, and no imagery data is delivered to the users.

This series architecture makes the *resilience of each and every element* vitally important, as its function is indispensable in delivering the capability. All key elements must deliver some minimum level of capability. Path redundancy at the element level is one common way to improve system resilience for these kinds of systems, making it more similar to the previous parallel case. For example, the post-processed data can be multicast to multiple satellites to ensure that at least one of the data links is available to any user.

### 1.3.3 Hybrid and Other Architectures

Not all architectures fall into the above two categories. A notable example is the Global Positioning System (GPS), which provides global positioning, navigation, and timing (PNT) services to millions of users. The performance of the GPS system, which in this case is the level of geolocation accuracy, is proportional to the number of satellites in view of the user's receiver. So, to a certain extent, the *performance* is proportional to the total number of satellites in the constellation. As the number of satellites is reduced, the performance degrades somewhat gracefully rather than resulting in an immediate outage. As with the parallel architecture, this highly distributed architecture provides its capability proportional to the number of satellites, however, in this case the capability of interest is the performance, which is the PNT accuracy obtained by the users.

Many architectures are a hybrid of both series and parallel elements. A series architecture can feature parallelism in the form of redundancy for some or all of the elements in the capability chain, as shown in Figure 1.4.

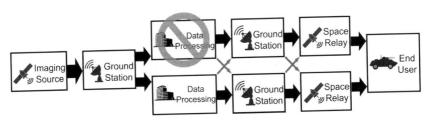

**FIGURE 1.4**
Example of a hybrid architecture.

In this way, some of the functions that are required to provide the end-to-end capability are provided in parallel fashion even though an end-to-end series structure is largely preserved. Cross-strapping between the parallel elements can provide even greater levels of redundancy and increased resilience, however, at some additional cost and complexity. In Figure 1.4, the loss of one data processing center does not result in the loss of end-to-end capability, as the second, redundant center is still available to process system data. This type of architecture has long been used to bolster system reliability by adding redundancy for elements that have lower reliability. However, redundancy of this type can similarly increase the resilience of such an architecture by improving the robustness for elements that are considered particularly vulnerable to one or more threats and thus make tempting targets.

## 1.4  Space System Elements

Each of the space system elements provides unique functions required to deliver the end-to-end capability. As a result, threats targeting individual elements have different impacts upon the system resilience. The following is a short overview of the system elements and their relevance to design for resilience.

### 1.4.1  The Space Segment

The space segment generally consists of one or more satellites orbiting the Earth. The key attributes of the space segment are the type, number, and orbit of each satellite, as well as other related features, such as inter-satellite crosslinks. Satellites can be large or small, depending upon the extent of their capabilities and role in the overall system. Commercial satellites vary in size but tend to be relatively large. Recently, "new space" companies have been trending toward larger constellations of smaller satellites. Most modern military satellites are large and complex and often support multiple missions. As a result of their high cost, there are usually a small number of satellites per system, although some systems, such as GPS, are composed of a larger number.

Satellites are operated from ground-based SOCs using radio links to send commands to each spacecraft. These commands can turn payloads on or off, switch to redundant units, point antennas, and otherwise perform routine housekeeping functions and payload reconfiguration. Information on the health and status of the satellite is returned to the ground station via telemetry links. Telemetry and command can also be accomplished using a gateway site once the satellite has been launched into its orbit and becomes operational. This is often accomplished using higher frequencies, a procedure called *in-band* telemetry and command in conjunction with the transmission of payload data.

The satellites usually transmit and/or receive payload data. This is information either generated by sensors on board the satellite or is relayed from another source. The earliest communications satellites acted as a "bent pipe" to seamlessly connect two distant parts of the world via radio signals. At that time data was limited to radio links carrying either analog voice and/or television signals. Today the signals and waveforms are more complex and the data include digital data files, text messages, or any number of other formats. Increasingly, these data links are operating at very high (wideband) data rates between 1 Mbps and 300 Mbps. These higher rates require larger antennas and higher transmit power, both on the satellite and at the ground stations. This is one reason that larger satellites have persisted.

A small number of satellites also host inter-satellite crosslinks. These are RF links between two satellites used to route data and/or signals. Military examples include the Milstar and Advanced EHF systems, which use crosslinks at a frequency of 60 GHz. A commercial example is the Iridium system, which uses 22 GHz crosslinks to route user data between satellites.

The use of crosslinks can reduce the number of ground gateways necessary to carry user traffic and can also enable ground sites to be placed only in very secure locations, or even eliminate the need for them altogether. This leads to a key trade for SATCOM systems: use of onboard processing versus ground processing. For systems in which data is processed and routed between source and destination, the processing can be performed either on the satellite in space or at a ground site. The ready availability of space and power on the ground has made it a preferred alternative, but as power efficiency and payload density increase, more and more digital processing can now be accomplished in space on the satellite.

### 1.4.1.1 Satellite Orbits

A satellite's orbit describes its position and motion relative to the Earth. The specific orbit selected is often based largely upon the system capability. Communications relay satellites benefit from geosynchronous orbits (GEO), while imaging satellites offer better performance at low Earth orbits (LEO). Weather satellites often are employed in sun-synchronous polar orbits. These different types of orbits are described by their altitude above the surface of the Earth and their angle relative to the equator (Figure 1.5). Each of these types of orbits has implications with regard to system resilience.

The lower the altitude, the faster the speed of a satellite relative to a point on the Earth's surface, making a satellite a more elusive target. However, the lower altitude also means a shorter range for any attack launched from the Earth's surface, requiring evasion techniques to be much more responsive. These trade-offs must be considered for certain threats.

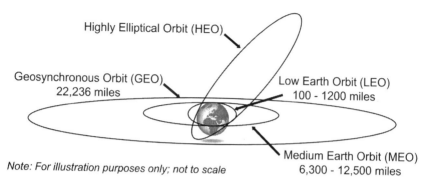

Note: For illustration purposes only; not to scale

**FIGURE 1.5**
Comparison of satellite orbits.

*Low Earth orbiting* satellites are in lower orbits (100 – 1200 miles) and are constantly moving with respect to any point on the Earth's surface. The exact orbital characteristics are determined by their mission. The inclination of a LEO orbit depends on the mission requirements. For example, some are placed into a polar orbit, which is good for many weather missions. These satellites orbit the Earth at an altitude of 300 to 500 miles, completing an orbit approximately every 90 minutes. Examples include the Defense Meteorological Satellite Program (DMSP) and the Polar Operational Environmental Satellites (POES), both of which deliver weather information to the U.S. government for military and civil uses. Ground stations that communicate with LEO satellites must possess tracking antennas that move with the satellites as they cross the sky to maintain the communications links.

Orbital debris is also a threat that can affect any orbiting body. There is more debris in the LEO orbital regime both due to greater congestion, probability of intercept between two moving bodies with relatively large velocity, and more human activity on platforms such as the International Space Station (ISS). Nevertheless, celestial objects such as asteroids and micrometeoroids are similar threats that can be difficult to track and can strike any satellite in any orbital regime.

*Medium Earth orbiting* (MEO) satellites are also moving with respect to the ground, but at a significantly higher altitude than LEO satellites, at between 6,300 and 12,500 miles altitude. GPS satellites, for example, are located approximately 12,550 miles above the Earth's surface, in a 12-hour orbit. Where LEO satellites might have a line-of-sight contact time of less than 15 minutes over any spot on the Earth, MEO satellites have a more persistent view. The commercial O3b satellite constellation also operates at MEO. As with LEO systems, MEO ground stations must employ tracking antennas. These satellites move more slowly with respect to the ground, but are at a much higher altitude, meaning any ground-based attack requires much longer transit times than for LEO satellites.

*Geostationary* (or *geosynchronous*) Equatorial *orbiting* satellites are located at approximately 22,236 miles above the Earth in an arc around the equator and appear to be motionless over a specific spot on the Earth due to their 24-hour orbit that moves synchronously with the Earth's rotation. As a result, these satellites provide persistent coverage over their field of view. Three satellites equally spaced can provide good coverage from 65°N to 65°S latitudes over the entire globe. This is the most popular orbit due to the advantages for communications satellites which can see approximately one-third of the Earth from an orbital slot positioned at a point over the equator. Many weather monitoring satellites also favor the GEO orbit, such as the Geostationary Operational Environmental Satellites (GOES), which provide most of the weather imagery for the continental United States.

These satellites are much farther from the Earth's surface than even the MEO satellites, and thus require even longer transit times for ground-based attacks. However, they maintain a known, stable position and are rarely moved, which makes them more vulnerable. This knowledge clearly favors an adversary targeting one or more satellites in GEO provided they have the means to do so. While physical threats in space are rare today, RF jamming is possible when the satellite's location is known.

*Highly elliptical orbit* (HEO) satellites are placed in highly elliptical orbits, usually with the nearest point occurring near the Earth's south pole, and the farthest point occurring above the far northern latitudes. A HEO satellite appears to linger over an area while at its farthest point (apogee) before increasing rapidly in relative velocity as it reaches its nearest point to the Earth (perigee). This allows the satellite to have a good view of the northern polar region of the Earth, and the elliptical orbit allows it to dwell there for much of its orbit to provide a long period of coverage. Russia (and previously the USSR) has been one of the most prolific users of the so-called *Molniya* orbit, which has an inclination, or angle, of 63.4° relative to the equator. HEO satellites often dwell at apogee at altitudes even farther than for GEOs, and then are moving very quickly as their range shortens. This combination of range and velocity imparts satellites in HEO some level of protection from ground-based threats versus those in some other orbits.

## 1.4.1.2 Satellite Composition and Size

A satellite is composed of two main components: the spacecraft bus and the payload. The bus provides the necessary functions for the satellite to operate, including a structure, and electrical power, thermal control, attitude control, telemetry and command, and propulsion subsystems. The payload provides the mission-specific functionality, such as sensing, navigation, or communications. The bus provides an infrastructure to enable the payload to operate as designed and communicates with it to also execute telemetry and command and data receive and transmit functions. Malfunction or incapacitation of either the bus or the payload can result in loss of the satellite capability.

Typically, the capability level of a satellite is proportional to its size. Larger satellites have more solar panels and larger batteries providing more power for the payloads and more fuel to enable longer mission life. A bigger bus also enables higher communications throughput and data rates by supporting higher power transmitters and larger antennas, and sometimes advanced onboard digital processing. For imaging and sensing satellites, larger satellites enable hosting of larger sensors with wider apertures and higher resolution, producing higher imaging quality.

### 1.4.1.3 Satellite Operating Frequencies

Another distinguishing characteristic of satellites occupying the space layer is their operating frequencies for space to ground communications links (uplinks and downlinks). The electromagnetic spectrum is categorized into a number of frequency bands. A frequency band is a contiguous span of frequencies allocated by national and international organizations, sometimes through treaties and agreements, for use by certain classes of users. U.S. satellite frequency allocations [3] include bands from ultrahigh frequencies (UHF) at 200 to 300 MHz to over 100 GHz for the millimeter-wave bands.

Operators of satellite systems must coordinate their frequency use through filings for the desired spectrum with such agencies as the International Telecommunications Union (ITU) and conform to national and international laws when using this spectrum. The availability of spectrum for certain popular frequency bands, particularly for GEO satellites, can present significant constraints in the design of a system. Frequency coordination activities can take months or years and must be considered early in the design process.

Figure 1.6 shows the most commonly used frequency bands for space systems. Certain bands are provided with letter designators, so the band at 8 GHz is referred to as X-band. Frequency bands are allocated to amateur, experimental, commercial, and government applications. For satellite operations, commercial bands exist at L-band (1.6 GHz), C-band (4 – 6 GHz), Ku-band (12 – 14 GHz), and commercial Ka-band (28 – 29 GHz). Government

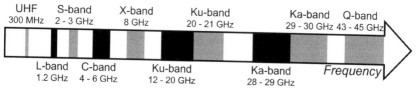

**FIGURE 1.6**
Satellite frequency band allocations.

bands occupy UHF (200 – 300 MHz), S-band (2 GHz), X-band (8 GHz), Ku-band (20 – 21 GHz), military (or federal) Ka-band (29 – 30 GHz) and Q-band (43 – 45 GHz).

Different frequency bands have unique characteristics and thus are used for different purposes. Satellite telemetry and command is usually performed at S-band, while data links occupy any of the other bands. Lower frequency bands, such as UHF and L-band, are narrower and provide less available bandwidth and thus capacity is limited. However, these bands are well suited to mobile users with smaller terminals and handsets. These bands are used for lower data rate applications such as voice. Higher frequency bands with more bandwidth more readily support higher data rate modes to include high-speed data transfer and high-definition video. While this is desirable for high capacity systems, higher frequencies are also more susceptible to signal loss due to weather conditions, such as heavy rainfall in a region, and thus their link availability is lower for a given signal power than lower frequency links. This is a trade-off that is made when designing the system. Over time the demand for more capacity has driven both the commercial and government satellite systems to higher frequencies. The other frequency bands between UHF and Q-band are reserved for other communications services, including terrestrial wireless.

The satellite design is driven by its operating frequencies. Antenna gain is inversely proportional to frequency and thus lower frequency operation requires larger antennas for a given sized coverage area and link QoS. An increase in transmitter power can compensate for smaller antennas to close the link and this represents a fundamental design trade. As satellite size and power are limited, space systems are often designed to reduce the communications burden on the satellite by shifting it to the ground stations. This is why many ground sites use high power transmitters and very large antennas. This is especially true for systems whose mission is not communications, as satellite size, weight, and power are generally reserved for the primary mission (sensing, navigation, etc.) rather than the communications payload or subsystem.

### 1.4.2 The Ground Segment

The ground segment is comprised of at least two key elements: the satellite command and control element, and the mission planning and operations element. In addition, there can be separate remote gateways to bring mission data to Earth as well as to data processing and dissemination centers. Many modern commercial and military SATCOM systems employ network protocols in the user terminals, allowing remote network control and interrogation. In this case, a NOC is required and this facility can be separate from the SOC.

While there is considerable concern today about the newfound perceived vulnerability of the space segment, the ground segment is arguably more at risk due to the increased amount of ground processing, the wider range of threats, larger number of dispersed facilities, and general accessibility of each. The ground segment for many systems has a large and conspicuous footprint. Military systems often publicly identify SOC and NOC locations, which are often secured on military bases on home soil, but many ground stations are distributed around the world, sometimes in remote and less protected locations.

Threats to ground infrastructure include physical threats, including sabotage, and cyber attacks. Networks often employ long-distance copper and fiber optic cables, which can be subject to both intentional and unintentional damage as well as eavesdropping. Gateways are deployed globally and thus are located in foreign countries (this is true both for the United States and other countries). These host countries might be allies but yet provide less security than locations on home soil. Interference and jamming can also be a serious threat for ground sites, as lower power transmitters can be located closer to the site and might also be mobile. For most systems the attack surface can be large. Cyber threats might be less pervasive in military computer networks but still represent an increasingly significant threat. Recent successful cyber attacks on U.S. civil space ground systems demonstrated the ability to inhibit operations and deny service [4].

### 1.4.2.1 Satellite Command and Control

Satellites must be operated and maintained while on orbit. They can drift in space due to gravitational effects, requiring regular station-keeping maneuvers to maintain their fixed position relative to the Earth (and other satellites). In addition, payload equipment such as antennas or telescopes must be commanded to point at desired locations, and occasionally operators must execute maneuvers or switch in redundant units in the case of on-orbit failures.

Satellite command and control is accomplished through RF links between the satellite and an Earth station, sometimes referred to as a telemetry, tracking, and command (TT&C) station. This ground station sends commands and the satellite returns telemetry that provides satellite status and state-of-health information. It can also accurately determine the satellite's position in space using the round-trip time delay between commands and the resultant telemetry, a technique known as *ranging*.

The satellite processes the commands which are then executed on board the spacecraft. Typically, these links are low data rates of less than 10 kbps and use the Space-Ground Link System (SGLS) or Unified S-band (USB) waveforms at S-band frequencies (~ 1.8 GHz) or "in-band" at other frequency bands being used for data transmission. Satellite status information is transmitted back to the ground via RF telemetry downlinks. These links provide health and status of the satellite and can also include other related information.

The actual generation of commands as well as receipt of spacecraft telemetry and ranging information generally does not occur at the TT&C station, but rather at the SOC, which may be located at a different geographic location. These two sites are connected by the terrestrial network.

### 1.4.2.2 Satellite Operations

Satellite operations encompasses all of the ground activities required to operate and maintain a satellite system. In most cases these activities are performed in the SOC, which is the source of satellite commands and the destination for satellite telemetry data. In certain unique systems, the satellite bus and payload are controlled by separate facilities, but this is much less common.

The amount of care and feeding that a satellite requires depends upon the complexity and function of the particular satellite. Some satellites are fairly static in their payload configurations and are rarely commanded for that purpose, with most commands used to maintain the satellite position or request data. Other satellites have advanced autonomy features, allowing them to operate independent of ground communications for an extended period of time.

The remainder of the satellites in orbit require more or less daily attention, and the loss of the SOC can severely limit the usefulness of the satellite system. For this reason, many systems feature both a primary and backup SOC. If operations are disabled at one location, the other can quickly assume operations. This also facilitates maintenance activities at each site.

### 1.4.2.3 Mission Planning

Mission planning is the act of creating a plan for satellite operation schedules that is executed in the SOC or in some cases a separate mission planning element (MPE). This plan scripts operations anywhere from hours to weeks in advance. User requests are received and processed to build the plan and to deconflict overlapping requests. Mission planning is a vital function for complex systems with many users and/or satellites and gateways. For SATCOM systems, resources such as bandwidth, coverage areas, and satellite transmit power are all functional constraints on the system that must often be actively managed by the operators. For sensing systems, mission planning includes conflict resolution for tasking by multiple users. The mission planning process allows that function to take place, ensuring that user requests can be satisfied with high confidence. If they cannot, the user receives a denial notification, explaining the reason for the denial, potentially allowing a new request to be made that can be satisfied.

In the past, a human satellite controller (SATCON) would manually execute these system commands, but now most systems are computerized, running algorithms to autonomously execute the plan. The complexity of the plan grows with the number of satellites and ground stations, the number of users, and the flexibility/dynamics of the system. Plans are often drafted days, weeks, or even months in advance and only updated when necessary. In many systems there is also an *ad hoc* process that can allow for high priority changes to the plan in near real time.

If the mission planning function is disabled, the capability of the system can be severely compromised, even if some period of time might pass before the effects are felt.

### 1.4.2.4 Gateways and Teleports

Most communications, imaging, and sensing systems use ground stations that provide gateways to terrestrial networks and transmit and receive data between the ground and the satellites. Certain military gateways are often referred to as *teleports* but provide largely the same function. Increasingly, these sites also feature equipment that converts the RF signals to digital formats for transmission across a network to the SOC or other destinations.

Satellite data is transmitted to ground gateways with large antennas (typically 6 – 18 meters in diameter) to receive high rate data from the satellites. For GEO satellites, each antenna is typically dedicated to a single satellite that is continuously transmitting and receiving data. In the case of LEO and MEO satellites, the gateway antenna must be capable of tracking the satellite as it moves overhead, and thus the antennas are shared among a number of satellites.

The geographic location of gateways is often determined by their field of view, which describes the gateway's ability to see the satellites that it must communicate with. The U.S. government maintains many gateways around the world to support many military and civil satellites in a variety of orbits. Gateways are often located such that each satellite is usually in view of at least two sites, such that if one goes offline for maintenance or due to a failure, the second site can be used for communications. The exceptions are LEO and MEO satellites which repeatedly pass over gateway locations and must exchange data in relatively short windows of access, in perhaps as little as 10 minutes. In these cases, the ground antenna must continually track the moving satellite for the duration of the contact time.

U.S. military gateways are well protected and located both in the United States and in allied countries, often on military bases. Nevertheless, as the loss of a gateway could reduce the effectiveness of military satellite systems, they represent yet another potential target by an adversary.

Commercial satellite service providers maintain their own gateways for communicating with their satellites. These sites can be located in their country of origin or in other countries around the world. Operations and maintenance of ground sites incurs significant expense so companies are motivated to minimize the number of ground sites to reduce operating costs.

Some companies' main business is the leasing of ground stations to customers for command and control, as well as data transmission. Sometimes these services act as risk mitigation against a failure of a satellite service provider's own ground sites. Generally, leasing of command and control services is a short-term option.

### 1.4.2.5 Network Operations

All of the ground elements discussed must be connected terrestrially. Today that is largely done using some combination of copper and fiber optic cables. The Department of Defense (DoD) uses both U.S.-owned and -leased lines to transport data. Both commercial and government users also employ the Internet in addition to dedicated leased lines. These lines interconnect the SOCs, mission planning elements, gateways, and any other data processing or storage facilities.

Once again, these connections represent a potential vulnerability. As with other system elements, path redundancy is the preferred solution to adding resilience to the network segment. Cyber security concerns also extend to the specific paths, owners of the lines, and geographic location, including geopolitical boundaries.

### 1.4.3 The User Segment

SATCOM systems support users who communicate with each other via user terminals. The first user terminals were simple radios that enabled low-rate data and voice communications through the satellite directly to other users. This is the simplest "single-hop" or mesh method of such communications modes.

As communications methods have advanced, more signal and waveform processing functions, as well as information assurance features such as encryption, have been added to the user terminals. While this provides greater performance and functionality, it also increases the cost and often the size, weight, and power requirement of the terminal.

User terminals can be fixed, transportable, or mobile. Military users often use either transportable or mobile terminals as operations move around the globe. True mobile terminals, operated while a platform is moving, are less common, although their use is increasing. Transportable and mobile terminals are less vulnerable to threats if their position is unknown and they move relatively frequently. This includes shipboard and airborne terminals.

The user terminals carry their own vulnerabilities to the system. If an adversary is able to capture a terminal, it could provide it access to the system, and potentially a means of penetrating the network as well as access to sensitive information. In addition, with newer digital equipment, eventually these nodes can expand the attack surface for a cyber threat. The terminals can also be used to disrupt uplink communications as they can be used as a jammer, providing unwanted interference when used improperly.

Taken together it is clear that there are many potential targets in a space system, each of which can cause a denial of service to its users. Each target is susceptible to a particular set of threats, some of which are common, while others are unique to the specific system element.

In the Introduction we noted that space systems are largely information systems and thus many of the credible threats are to the links that interconnect the system elements and convey the data. The destruction of a system node is not necessarily required to inhibit system traffic, nor must the threat produce irreversible damage nor even a long-term adverse effect to accomplish its goal.

The more that is known about the system, its elements, and its interfaces and links, the more likely it is that the credible threats are recognized, allowing for effective mitigation features to be incorporated into the system design.

# 2

## Defining and Evaluating Resilience

Though a universal approach to defining, measuring, calculating, and evaluating resilience has yet to be widely accepted, certain publications in the literature have captured the key concepts that can be employed in the system design process. A key reference for this book is one of the first published definitions of resilience for space systems that can be found in the U.S. Department of Defense (DoD) Fact Sheet, "Resilience of Space Capabilities," published in 2011 [5] (Figure 2.1).

> *"Resilience is the ability of an architecture to support the functions necessary for mission success in spite of hostile action or adverse conditions. An architecture is 'more resilient' if it can provide these functions with higher probability, shorter periods of reduced capability, and across a wider range of scenarios, conditions, and threats. Resilience may leverage cross-domain or alternative government, commercial, or international capabilities."*

At its most fundamental level, resilience is the property of a system maintaining its core capabilities in the presence of one or more external threats that can interrupt its operation, denying services to its users. This definition is useful as a foundation for space system design, which focuses on developing a system solution that delivers one or more capabilities to the operator or user regardless of threats.

This definition provides a thorough description of resilience as applied to space systems, albeit with an emphasis on military applications. It further includes a basis for measuring and quantifying resilience for space systems by including key definitions for an appraisal of resilience:

- It is clearly a capability-based attribute: The criteria are "mission success," "providing functions," and "capability."
- It is applied specifically to the situations involving "hostile action" and "adverse conditions"—the attribute is based upon specific threats.
- Value criteria are provided such that greater resilience is defined by providing functions with "higher probability," and "shorter periods of reduced capability," "across a wider range of scenarios, conditions, and threats."

## FACT SHEET: Resilience of Space Capabilities

*As we invest in next generation space capabilities and fill gaps in current capabilities, we will include resilience as a key criterion in evaluating alternative architectures.*

*National Security Space Strategy*

The National Security Space Strategy (NSSS) charts a path for the next decade to to maintain and enhance the advantages derived from space while confronting the challenges of an evolving space environment The NSSS seeks to address a strategic space environment that is increasingly congested with increasing amounts of space debris, contested by a growing range of foreign counterspace capabilities; and competitive as more and more countries and companies operate in space. Resilience is one way to address this more challenging space environment. The strategy notes that strengthening the resilience of our architectures can deny the benefits of an attack on our space infrastructure, as well as enable our ability to operate in a degraded space environment.

### Key Ideas Underpinning Resilience

*The purpose of resilience is to assure performance of military and related intelligence functions at a level necessary to execute assigned missions within an acceptable tolerance for risk.* This functional mission assurance must account for the full range of anticipated scenarios, conditions, and threats that drive our planning. Combatant Commanders largely define "acceptable risk" for military functions in consultation with the Secretary of Defense, Director of National Intelligence, and Commander in Chief

*We primarily seek to make resilient the military functions dominantly provided by space systems.* Thus, the focus is on traditional missions that support the warfighter, as well as the underlying missions required to conduct space operations.

*Resilience is comprised by capabilities from multiple domains.* Therefore, resilience is evaluated at the enterprise, mission, or functional level in a manner that encompasses the systems and system of systems provided by multiple domains that enable a given function. The evaluation of resilience must also comprehend the contributions of capabilities within a domain.

*Resilience to both hostile actions and adverse conditions is needed.* Therefore, resilience must equally consider threat-based hostile acts, as well as aberrations caused by any number of natural or man-made adversaries.

*Resilience focuses on maintaining or replenishing capabilities, and thus transcends conventional risk mitigation efforts.* Risk management and mitigation initiatives primarily focus on reducing threats to components and systems. By focusing on sustaining critical capabilities, resilience shifts thinking from the protection of key assets to the sustainment of key capabilities and the maintenance of the functional enablers that support these capabilities.

*We must strike a balance between risk-based functional performance, resilience, and affordability.* Resilience may not always require increased investment. Changes in policy, practice, or procedure can offer real operational value. Cost sharing with allies, as well as leveraging commercial hosting opportunities, can add performance and resilience.

Resilience encompasses *avoidance, robustness, reconstitution, and recovery*

- **Avoidance:** countermeasures against potential adversaries, proactive and reactive defensive measures taken to diminish the likelihood and consequence of hostile acts or adverse conditions
- **Robustness:** architectural properties and system of systems design features to enhance survivability and resist functional degradation
- **Reconstitution:** plans and operations to replenish lost or diminished functions to an acceptable level for a particular mission, operation, or contingency
- **Recovery:** program execution and space support operations to re-establish full operational capability and capacity for the full range of missions, operations, or contingencies

### Definition

*Resilience is the ability of an architecture to support the functions necessary for mission success in spite of hostile action or adverse conditions. An architecture is "more resilient" if it can provide these functions with higher probability, shorter periods of reduced capability, and across a wider range of scenarios, conditions, and threats. Resilience may leverage cross-domain or alternative government, commercial, or international capabilities.*

### Levels of Evaluation

Resilience can be measured at multiple levels. The primary measure of resilience at the Enterprise, Mission, and Functional levels is risk to national security objectives, mission effectiveness, or functional capability. Resilience is also assessed at the Domain, Constellation, and individual Space System level.

### Criteria for Evaluation

The five evaluation criteria below provide a common measure to assess resilience for any given functional architecture.

1. Anticipated *level of adversity*
2. *Functional capability goals* necessary to support the mission
3. The *risk* that these goals may not be met at a given level of adversity
4. The *severity* of the functional shortfall to the mission
5. The *time* which the shortfall can be tolerated by the mission

The temporal component of this evaluation construct is of particular import. Time primarily quantifies the reconstitution component of resilience. Resilient is presumed to be a more lengthy replacement effort tied to more traditional planning and programmatic activities that extend well past the period of crises.

### Next Steps

This resilience definition and criteria can form a basis from which to institutionalize resilience into our architectures, requirements, planning, programming, acquisition, and operations activities. Resilience is essential to assure our functional capabilities, particularly those enabled by space, at a time when the domain is increasingly congested, contested, and competitive.

**FIGURE 2.1**

2011 DoD Fact Sheet on space resilience.

This definition is important as designers must assign value to the various contributions to the resilience of the system and eventually quantify them in a meaningful way that reflects the value of uninterrupted service to the user. The user ultimately does not care how the service is provided, simply that it is delivered with some minimum quality of service (QoS). This provides flexibility to the system designer when weighing how best to meet the resilience requirements. In addition, though this Fact Sheet was published by the U.S. DoD, the definition is actually very broad and applicable to a wide range of space systems with minimal tailoring.

## 2.1 Resilience Domains, Attributes, Timeline, and Criteria

The Fact Sheet also states that *"Resilience is comprised by capabilities from multiple domains. Therefore, resilience is evaluated at the enterprise, mission, or functional level in a manner that encompasses the systems and system of systems provided by multiple domains that enable a given function."* This means that the highest architectural level is the "System of Systems," often referred to as the "enterprise" level. The space system is part of the enterprise and is comprised of segments. Each segment could be further comprised of individual elements. An example of such a hierarchy is shown in Figure 2.2.

This introduces the concept of global and local (or elemental) resilience, which becomes an important trade. A system element might be resilient to a localized threat, however, it could be but one contributor to the resilience

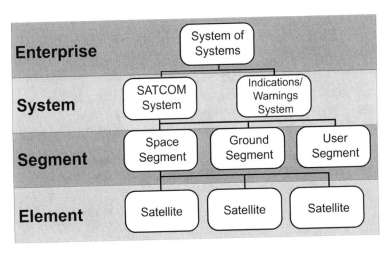

**FIGURE 2.2**
Example of a hierarchical system model.

of the greater system, as that element represents only a portion of the total system mission capability. For example, one satellite in a constellation might not be resilient to a threat, but the loss of that satellite's capability represents a negligible loss of total mission capability and thus represents a small impact on the overall system resilience. This concept of system or enterprise level resilience versus elemental resilience is among the key design trades described in Chapter 6.

The source Fact Sheet identifies and defines four system attributes that provide resilience:

1. **Avoidance:** Countermeasures against potential adversaries, proactive and reactive defensive measures taken to diminish the likelihood and consequence of hostile acts or adverse conditions
2. **Robustness:** Architectural properties and system of systems design features to enhance survivability and resist functional degradation
3. **Recovery** (called "Reconstitution" in the Fact Sheet): Plans and operations to replenish lost or diminished functions to an acceptable level for a particular mission, operation, or contingency
4. **Reconstitution** (called "Recovery" in the Fact Sheet): Program execution and space support operations to reestablish full operational capability and capacity for the full range of missions, operations, or contingencies.

*Note:* Due to heritage space industry standard conventions, from this point forward the "recovery" and "reconstitution" definitions have been interchanged. *Recovery* refers to single-mission, shorter-term reestablishment of capability and *reconstitution* refers to full replacement restoring capability for all missions. This is done to maintain consistency with the established space industry lexicon and does not materially impact the methodology presented in subsequent chapters.

The effects of the four resilience attributes occur sequentially in time, as shown in Figure 2.3. Each attribute can be viewed as a measure of the

**FIGURE 2.3**
Sequence of resilience attributes.

system's ability to sustain the effects of a threat and maintain some level of operation. The first opportunity for achieving resilience is avoidance. For a threat to be avoided, there must be a means to detect and identify it; these functions may be considered part of the system or part of another system that provides these support functions. If the threat is *fully* avoided, then no capability is lost. If the threat is not fully avoided, the extent to which capability is lost then depends upon the robustness of the system to the threat. If some capability is lost, the system might possess the ability to recover some or all of the lost capability after some period of time. Finally, the non-recovered capability could be completely reconstituted, restoring capabilities for all missions or services.

## 2.2 Valuing Resilience

Implicit in the definition, and central to the system design process, is the need to quantify, in some way, the resilience of a system. This means creating a method of calculating a value for the system's response to a threat. The more likely the system can avoid a threat's impact, the greater the value of its resilience metric. Likewise, greater robustness equates to reduced capability loss and higher resilience. In the cases of recovery and reconstitution, the value of capability versus time comes into play, as the utility value can vary depending upon the durations of the outage time. All of these effects and attributes should be considered, valued, and combined to create a quantitative assessment of the overall resilience of a system to a threat.

System utility is often a time-dependent quantity. The military often speaks of "tactically relevant operational timelines," meaning that if a resource is not available within a certain time period, its value to the warfighter is negligible, if not zero. Likewise, service outage time in the commercial satellite service industry corresponds to lost revenue and customer dissatisfaction. So, in addition to the minimum capability required, the duration over which it is lost is also an important factor when calculating the resilience value.

The criteria for evaluating resilience are also provided in the DoD Fact Sheet:

1. Anticipated level of adversity
2. Functional capability goals necessary to support the mission
3. The risk that these goals may not be met at a given level of adversity
4. The severity of the functional shortfall to the mission
5. The time which the shortfall can be tolerated by the mission

It is highly significant that these criteria assert the need to establish a minimum mission capability ("functional capability goals," "severity of shortfall," "time of shortfall") necessary to adequately support the mission. This is addressed extensively in later chapters.

## 2.3 Prerequisites for Evaluating Resilience

To evaluate different system designs, a resilience analysis is required, evaluating each design against the validated threats. The value of resilience analysis is to determine what, if any, impact an event generated by a threat has on the system. Simply identifying the threat does not make any assumptions as to the likelihood that the threat is realized. The question is how well the system is expected to survive and maintain its services given that the threat is initiated.

Since the result of a resilience analysis is a prediction of the behavior of the system, it is also important to clearly define the meaning of the result. A simple methodology is presented in Chapter 5 that enables a quantitative calculation of resilience. The obtained resilience values have a specific meaning and must also be compared against the minimum acceptable value. Three key pieces of information are required to determine the system resilience: the system capability, the threat(s), and the concept of operations (CONOPS) (Figure 2.4). The greater the insight that the system designer has into each of these aspects, the more likely a resilient design can be created.

1. System Capability Definition

   It is essential to describe the system capability and associated performance requirements as accurately and completely as possible. Resilience has little meaning without an understanding of what is required of the system.

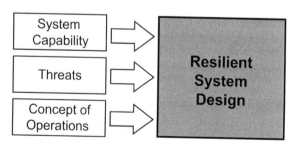

**FIGURE 2.4**
Inputs to the resilient system design process.

Questions to be answered include:

- What is the system capability being delivered?
- How does the system deliver that capability (system architecture)?
- What is the minimum system capability value required in contested or stressed conditions?

Answers to these questions help determine whether or not sufficient resilience exists to meet user needs. Though resilience can be calculated in the absence of such a criterion, from a designer's point of view, a requirement should be established to drive the trades between resilience, performance, and cost.

2. Threat Definition

It is equally important to describe the threats as accurately and completely as possible prior to designing resilience into a system. Though the system designer never has as much information about the threat as desired, having as accurate a definition as possible will improve the accuracy of the resilience estimate and the utility of cost, performance, and resilience trades.

Questions related to threat definition include:

- What is the nature of the threat: physical, electronic, cyber, optical?
- What is the severity of the threat?
- What is the threat's estimated effectiveness?
- What is the threat's persistence?
- What are the threat's targets?

The specifics of threat definition are presented in Chapter 3.

3. Concept of Operations (CONOPS) Definition

The manner in which the system is operated is called the concept of operations (CONOPS), and it can affect the resilience of the system. Operators of space systems may have threat mitigation options that are based upon the ability to change or reconfigure the system's state. These actions can be performed by human operators or autonomously. For example, communications systems might have path diversity in routing data from source to destination. Including more options in the operation of the entire end-to-end system provides flexibility in dealing with certain threats. And the speed at which human operators can execute responsive actions when a threat is identified can impact the depth and duration of a capability outage.

Collectively, system capability, threat, and CONOPS information is used to inform the design, resulting in a system with the desired features and performance. Once these three pieces of information come together to form a conceptual design with a defined implementation, the resilience analysis phase begins.

Designing for resilience is simply a part of the traditional systems engineering design process. There are rarely any silver bullets that provide all of the required resilience. Rather, there are usually a number of threat mitigation techniques and tools that the designers have at their disposal which, when combined, can provide the required resilience. This could entail using both novel system architectures and selected local protection features to achieve the resilience goals. This trade is discussed in greater detail in Chapter 6.

## 2.4 Approaches to Calculating Resilience

From the Fact Sheet definitions several key elements of a resilience calculation methodology are clear. The resilience metric will be a function of four system attributes: Avoidance, Robustness, Recovery, and Reconstitution. There are potentially many ways to approach this calculation, including a number already presented in published literature. Before proceeding, a brief discussion of different approaches for considering threat probabilities and avoidance is warranted to examine the pros and cons of each.

Once a clear definition of resilience has been established, including system attributes and evaluation criteria, a method of using this information to calculate the resilience of a space system can be developed. Ideally the resilience metric should have some recognizable meaning related to the predicted system performance following the exercise of a threat.

At this point it seems clear that any useful resilience metric should reflect the predicted loss of system capability for a particular threat. This loss of capability should include both depth and duration of the loss. It should also include the complete expected system behavior when the threat is encountered, to include such functions as recovery.

Perhaps the simplest approach can be formulated in which the resilience is a calculation based on the depth of capability loss combined with the duration of that loss over some period of interest. Figure 2.5 provides an example of a normalized value of resilience that is the integral of the capability loss under the curve versus time. As in most examples in this book, the system capability has been normalized to 100 percent (or 1.0). This approach could be sufficient if avoidance is not part of the calculation, but a more complete approach is desirable.

Once avoidance is included, several subtleties must be addressed. Perhaps the single most debatable point relates to the interpretation of the statement that a space system is more resilient if it "provides these functions with higher probability." This has been interpreted in at least two different ways and hinges upon

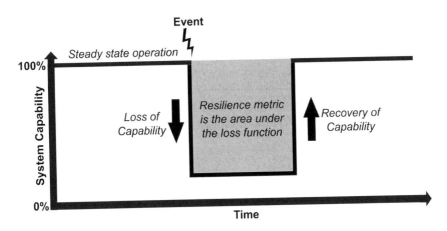

**FIGURE 2.5**
Simplified resilience calculation profile.

whether or not the *probability of the threat occurring* should be part of the actual system resilience calculation, which is not explicitly included in the definition.

The rationale for including the probability of threat occurrence is that these probabilities scale the impact to the system over a threat portfolio, essentially pro-rating the loss of capability by relative likelihood of the threat. This has an advantage of making a single calculation easier to obtain across multiple threats. This approach also shares certain similarities with risk management techniques in which risks are ranked using two scales, for "likelihood" and "consequence," each with a value of 1 through 5 (Figure 2.6). Higher values indicate increased likelihood or consequence. Likelihood is the probability of the threat occurring, while consequence is the assessment of the impact to the system capability if the threat is manifested. Typically, the two values are

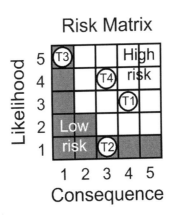

**FIGURE 2.6**
Risk matrix for threat assessment.

multiplied to provide a "risk metric," Risk = L × C. For L = 4 and C = 3, the resulting risk value is Risk = 4 × 3 = 12.

Areas of the risk matrix with very low likelihood and/or consequence represent areas of low risk, and so if a threat falls in that region it could be deliberately excluded from the design process. Areas of the matrix with high values for one or both represent areas of high risk and threats falling in that region should be considered. In Figure 2.6, threats T1 and T4 are in the high-risk region, whereas T2 and T3 are in the low-risk region. This is discussed in more detail in the system design context in Chapter 6.

From a resilience perspective, the "consequence" part is the actual resilience of the system to the threat. The "likelihood" part is the probability of the threat occurring. Taken together, the designer can use this assessment to determine how to best allocate costs, assigning greater resources against threats that have both high likelihood and consequence (low resilience). In practice, the early screening of threats as part of the system requirements definition performs the same function and is separable from the basic resilience calculation.

However, from a designer's point of view this approach is not a particularly useful definition because a key parameter in the resilience calculation, the probability of the threat occurring, is beyond the control of the designer. As a consequence, there is no threat mitigation approach or feature that can affect this probability. Instead, knowledge of this value can be used to guide the designers in trades between cost and resilience, influencing the development of the requirements as to which threats to include in the design process. This preserves the evaluation of resilience as strictly a behavioral value.

This definition also implies that the system's ability to withstand the impact of some threat is dependent in some way upon the probability of that threat actually occurring. As the composition of the system is what provides its capabilities, including avoidance, robustness, recovery, and reconstitution, the idea that the *probability* of an event's occurrence modifies the system behavior is counterintuitive from an engineering point of view, since it does not change the system. What is true is that the chosen resilience metric is higher for lower probability threats, which is just another way of saying that the system is less likely to suffer loss of capability simply because it is less likely to be impacted. That part is clear but does not provide objective insight into the system's behavior versus the threats.

That said, this resilience metric is as valid as several other proposed methods. However, the calculation is more complicated, and it does not easily lend itself to aiding system designers in executing trades and implementing threat mitigations. And determining the likelihood of a threat occurring is one of the least accurate assessments to be made. Still, this approach has similarities to the one that is presented here; there is more similarity than difference. However, the approach in this book is intended to better serve design activities through a simplified methodology using parameters that directly relate solely to a system's design.

In the methodology that follows, it is noted where and when the probability of a threat occurring is useful, but strictly speaking it is not included as a part of the basic resilience calculation. The use of the risk management approach to evaluate and screen threats is a separate step that occurs prior to any calculation of resilience and results in the threat requirements that are then presented to the system designer. It is necessary to decide which of the many identified threats are worth considering in the system design. But once the threat range is narrowed by this filtering step, the probability of occurrence is not integral to the general capstone calculation presented in Chapter 5.

Instead, there is another probabilistic parameter, that of the *probability of threat avoidance*. This is the alternate interpretation of the definition of providing functionality with higher probability. This is a different parameter, describing the probability that the threat in question is fully avoided once it has been realized. The greater this value (which varies from 0 to 1), the higher the resilience, which represents a higher probability of delivering system functionality. This approach results in a slightly different resilience metric: *the expected value of the residual system capability when exposed to a threat.* It represents the predicted response of the system given a specific threat. How likely that the threat is carried out is not material to the calculation; resilience is predicated upon a threat being realized and then determining how the system capability is impacted. This is not to imply that likelihood does not have a place in the final trade, but simply that it is not relevant to the individual calculations. It is not a question of whether the adversary presses the "big red button" but what the response of the system is once that button is pushed.

## 2.5 Resilience Calculation Parameters

The capability-based timeline shown in Figure 2.3 forms the starting point for the development of a simple behavioral system model representing the system response to a threat. The input to the system model is the initiation of an action due to a detected and identified external threat, and the output is the expected residual capability metric, a resilience value, based on the predicted system behavior. This is illustrated in the notional simplified system behavioral timeline shown in Figure 2.7, building on the similar profile presented earlier in Figure 2.5. In this figure, if the threat is avoided there is no loss of capability. If the threat is not avoided, a significant loss of capability occurs. At some later point in time, a portion of that capability is recovered. Later still, full reconstitution of the initial capability is attained. There is no assurance that full capability can be recovered in many instances, but that is a goal of the designer.

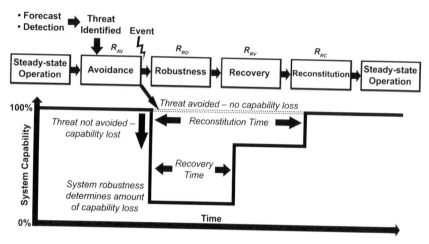

**FIGURE 2.7**
Capability-based resilience timeline.

This timeline describes the system response in terms of its resilience attributes, representing its ability to cope with the threat. It is then possible to estimate the expected value of the retained capability following the exposure to the threat. The resilience value R will take on a value between 0 and 1, indicating the portion of the initial capability that is retained after all threat mitigations have occurred. This estimate is derived from an equation containing the four variables shown in Figure 2.7: $R_{AV}$, $R_{RO}$, $R_{RV}$, and $R_{RC}$. Note that each of the resilience attributes, discussed earlier in this chapter, has been parameterized. Each of these is a variable for the related attribute and is assigned a value between 0 and 1. The avoidance metric, $R_{AV}$, is unique in that it is by definition probabilistic and represents the *probability of avoidance*. The other three variables are scalar fractions; for example, $R_{RO}$ is the fraction of capability projected to be retained if the threat is not avoided and the system is impacted. The detailed mathematical models are discussed in Chapter 5.

The utility of using the resilience attributes as resilience parameters lies in the fact that each of these coefficients can be directly related to specific categories of threat mitigation approaches. Avoidance can be obtained through the use of techniques that cause the system to avoid the impact of the threat altogether, such that the event never occurs. Robustness can be obtained either through the choice of system architecture, or via the use of specific protection features throughout the system. Recovery, whether manual or autonomous, is a proven method for restoring system capability. And reconstitution is a wholesale replacement of one or more parts of a system to restore service. Figure 2.8 shows examples of various techniques for each that are discussed in greater detail in later chapters.

| Avoidance | Robustness | Recovery | Reconstitution |
|---|---|---|---|
| • Countermeasures<br>• Deterrence<br>• Mobility<br>• Maneuverability<br>• Covertness | • Active redundancy<br>• Overcapacity<br>• Excess margin<br>• High damage<br>thresholds | • Passive redundancy<br>• Repair<br>• Reset / restart<br>• Self-healing<br>• Threat neutralization<br>• Autonomy | • Replace<br>• Rebuild |

**Resilience**

**FIGURE 2.8**
Methods of imparting system resilience.

Source: "A Method for Calculation of the Resilience of a Space System," MILCOM 2013,
R. Burch, 2013.

A detailed discussion of the derivation of this resilience quantification methodology is presented in Chapter 5. While this specific approach is presented, there are a number of other, similar approaches that can be found in the literature. The majority of these approaches are based on the same source material and differ in the level of mathematical rigor or interpretation of the definitions. However, the basic approach to quantifying resilience remains fundamentally similar, as examples show. In some cases, these alternate methods are special cases of the more general methods presented in this book.

## 2.6 The OSD Taxonomy of Resilience

In 2015, the U.S. Office of the Assistant Secretary of Defense (OSD) published a white paper describing a "taxonomy" of resilience [6]. This is notable because it is among the first attempts to standardize the vocabulary of resilience for the space domain. This paper specifies six "sub-elements of resilience": distribution, disaggregation, diversification, proliferation, protection, and deception (Figure 2.9).

These six sub-elements differ from resilience attributes in that they are methods of incorporating the attributes to impart resilience to the system. It is useful to consider each of the six in the context of the types of impact that they have on the system design. One seeks to protect the system from harm, four in some way attempt to either avoid or limit the loss of capability in the case of a successful event, and one seeks to completely avoid an event altogether. It is important to consider the implications of each sub-element and *how* it imparts resilience to the system.

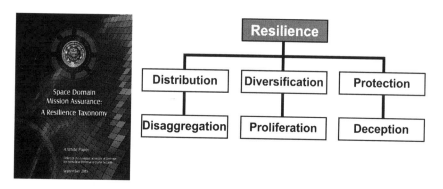

**FIGURE 2.9**
OSD SDMA sub-elements of resilience.

Source: "Space Domain Mission Assurance: A Resilience Taxonomy," Office of the Assistant Secretary of Defense for Homeland Defense & Global Security, 2015.

The attributes express how the system behaves, and this is discussed mathematically in Chapter 5. In contrast, the sub-elements represent methods that result in one or more of the attributes being present in the system design. Figure 2.10 shows a notional mapping between the two. The sub-elements are then expressed in the design as specific functions or features included in the implementation of the system. Note that though OSD chose these six sub-elements, they do not necessarily include all of the means of providing resilience to a system. For example, reconstitution was described as a different branch at the same level as "resilience" in the taxonomy, under the higher category of "space domain mission assurance."

However, each of these six sub-elements should provide one or more of the remaining three attributes described earlier (avoidance, robustness, recovery). From these general categories, specific mitigation techniques can then be derived. Several of these sub-attributes are used in subsequent examples of resilient architectural options presented in later chapters.

| Resilience Sub-Element | Resilience Attributes | | | |
|---|---|---|---|---|
| | Avoidance | Robustness | Recovery | Reconstitution |
| Protection | X | X | | |
| Deception | X | | | OSD lexicon does not include reconstitution as part of resilience |
| Distribution | | X | | |
| Disaggregation | X | X | | |
| Diversification | X | X | X | |
| Proliferation | | X | | |

**FIGURE 2.10**
OSD sub-elements mapped to resilience attributes.

## 2.7 Other Resilience Nomenclature

This book begins with first principles and then applies them to the design of space systems. This is not to imply that the topic is fully and equally addressed. Related disciplines have already emerged to deal with the topic of resilience in ways more suited to each specific application. A notable example is cyber resilience, which is a more specialized area with its own unique lexicon and unique threats and mitigations, representing a discipline unto itself. As cyber attacks represent a credible threat to space systems, the topic is discussed in this text but only to a limited degree. Literature containing a more comprehensive presentation of cyber resilience can be found in the References section at the end of the book.

# 3

## Threats

Threats represent the primary external factor driving the design of a resilient space system. The more accurately these threats can be defined, identified, characterized, and understood, the more likely the threat mitigations devised and incorporated into the design successfully preserve the system capabilities in a hostile or hazardous environment.

Determining which of these threats can interrupt system performance is crucial in creating a system that is optimally designed with respect to cost, performance, and resilience. This chapter introduces and discusses several of the most common threats and their properties. The associated mitigation approaches are discussed in Chapter 4.

## 3.1 Categorizing Threats: Adverse Conditions and Hostile Actions

The dictionary definition of a threat is "an expression of intention to inflict evil, injury, or damage," or "an indication of something impending." Both are appropriate as applied to the study of threats against a space system. Threats are actions and conditions that can result in events that interfere with a space system's ability to deliver its intended capabilities. A threat must be considered *credible* for it to be included in the system requirements. Only credible threats are deemed worthy of investment to counter or mitigate their potential effects to the system. A threat assessment is performed to identify credible threats—this will be discussed in Chapter 6. The threat might not exist in the present but is projected to exist in the future over the life of the system.

In categorizing the types of threats, the first distinction is between adverse conditions and hostile actions (Figure 3.1). The first definition above implies a human hand in the creation of the threatening condition, the threat of a *hostile action*. The second definition, "an indication of something impending," is more descriptive of a threat representing an *adverse condition*, one that is not due to human intent. Sometimes these are also referred to as *hazards*.

A *threat scenario* is a situation that includes one or more threats and/or adverse conditions occurring in some combination or sequence. One threat could be an adversary jamming a satellite receive uplink. A second threat

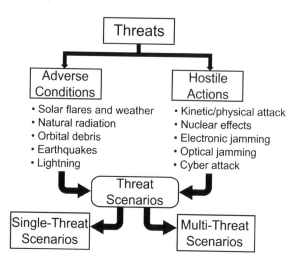

**FIGURE 3.1**
Threat categorization.

Source: "A Method for Calculation of the Resilience of a Space System," MILCOM 2013, R. Burch, 2013.

could be a physical attack on a ground station. A threat scenario therefore encompasses the two threats executed in a coordinated or sequential fashion. The scenario is often expressed as a timeline of events that describes the projected deployment of the threat(s) and the related conditions. These scenarios are meant to be representative of those that the system must sustain during operations and serves as a method of evaluating resilience against specific predefined cases.

### 3.1.1 Adverse Conditions

Adverse conditions are *not* the result of intentional human actions and include terrestrial and space environmental conditions that can interrupt system operations. Examples include severe weather, "acts of God" such as earthquakes, or a meteoroid or piece of orbital debris striking a satellite. These are often treated as random events which can be anticipated, but the exact timing and severity are unknown. In space, space weather due to solar activity is a concern, including coronal discharge, solar flares, and other phenomena. Additionally, there is the growing threat of a conjunction between a spacecraft and orbital debris. Severe terrestrial weather can also impact RF communications links between the satellite and the ground station and users. In some instances, there is a warning, permitting threat mitigation actions to be activated in an attempt to avoid the threat, while other times there is not.

There are also some threats that are the result of human actions without malice and are thus considered more of an adverse condition than a hostile action. An example is the unintentional interference to a satellite caused by a mispointed ground terminal. In this case human error is to blame, but this is not considered a hostile action with intent to disrupt service. Nevertheless, the impact upon the system is the same, so the threat remains.

### 3.1.2 Hostile Actions

In contrast to adverse conditions, hostile actions are threats that are the result of human intent with the specific goal of disrupting the operations of the space system. Intentional RF jamming of satellite receivers, anti-satellite weapons (ASATs), and a nuclear detonation in space are all examples of hostile actions. It is worth noting that many, if not all, of the threats to space systems of concern today have been described since the 1980s [7]. The difference today is the level of technological development that now makes many of these more achievable and thus credible.

An adversary's intellect and sophistication can result in highly targeted threats based on its knowledge of the system that it wishes to disrupt. More capable adversaries can produce more severe threats with a higher probability of success. While the exact timing of hostile actions can be difficult to predict, their appearance may be anticipated based on careful observation of an adversary's behavior and the use of surveillance systems or intelligence that can provide early warning of an imminent attack. These types of systems provide *indications and warnings* (I&W) which can be inputs into the space system that are used to activate mitigation features.

An adversary must possess the *motivation, means,* and *opportunity* to pose a threat to a space system. There is *motivation* when there is an incentive for attempting to disrupt the system because the benefits outweigh the costs. The *means* is possession of the capability to actually disrupt the system function assuming that the threat is successfully activated. The *opportunity* is the ability to actually exercise the inherent capability that manifests the threat. To preempt a threat at the source, at least one of these three necessary prerequisites must be removed.

The primary goal in designing for resilience is in determining how the system is expected to respond to a threat that is *realized*, meaning that it is manifested as an *action* which can result in an *event* (Figure 3.2). For the purposes

| Decision to Act | Initiation of Action | Action in Progress | Event Occurs |

**FIGURE 3.2**
Hostile action sequence: Decision to event.

of illustration, an event is anything that has the potential to adversely change the state of the system. The event might result in an *impact* to the system, depending upon the threat effectiveness and the effectiveness of any system features for threat avoidance and robustness. In the case of a threat successfully avoided, the event can be considered more of an encounter which triggers a system response unless the attempt to realize the threat completely fails. In the case of a failed attempt, the system is not impacted and there is no state change. However, if in avoiding a threat the system is caused to lose some capability for a period of time then the system has less resilience even though it was not directly impacted.

It is useful to consider a specific example to help illustrate the threat lexicon. The simple example in Figure 3.2 includes all of the potential elements of interest in the timeline. The *a priori* knowledge that a direct ascent ASAT launch capability is in place coupled with one or more observations of demonstrated capability, such as a successful test launch, would be enough to establish such an adversary's capability as a credible threat. There is established motive and means. When the adversary decides to execute a threat, a space system is targeted, an action is initiated, and a launch *attempt* occurs. If the attempt fails (which is a possibility), no event occurs that affects the space system. If successful, there is some amount of time during which the missile is in flight as the action progresses. Upon approaching its target, the missile might still miss. If the threat action was not sufficient to trigger a system response and no change in state to the system occurs, it could be argued that no event occurred. However, if there is a detection and a system response, an event has definitely occurred, regardless of its ultimate impact upon system operation. Likewise, if the missile impacts the target, independent of whether or not it causes damage, an event has clearly occurred.

Accurate threat definition is crucial in designing a resilient system. The more accurately the threat is characterized, the more likely the system can be designed to counter it by minimizing its impact. For most systems there is more than one credible threat, leading to the establishment of a threat range defined by the individual threats in the threat portfolio. This is illustrated in Figure 3.3.

Anticipating the nature and severity of threats is vital in designing for resilience. If the threat is underestimated, the system will be designed with insufficient resilience. If overestimated, the system will be overdesigned with the additional cost providing little or no value and potentially compromising performance.

Worst-case conditions are usually assumed when evaluating the system resilience, as with traditional engineering design approaches. In the case of threats due to hostile actions, this means assuming that the adversary has a certain level of sophistication and knowledge based on its perceived capability in posing the threat. An adversary's knowledge of the design characteristics and features of the system can enable crafting of a more effective strategy in exploiting specific system vulnerabilities. As a result, specific

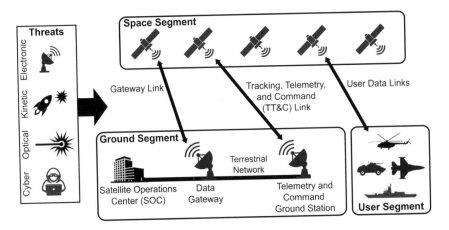

**FIGURE 3.3**
Space systems can be affected by a wide range of threats.

performance specifications and design details are often kept secret so as not to aid an adversary in its development of threats to the system. The consequences of threat severity based on level of adversary sophistication is discussed in Chapter 6.

## 3.2 Threat Attributes and Characteristics

Defining threats requires a description of the attributes such that their behavior and potential system impacts are known and understood. These attributes include their type (or nature), severity, target(s), and effectiveness. Some of this information is based on the source of the threat, including the nature of an adversary considering a hostile action.

### 3.2.1 Threat Types

The first step in defining a threat is to consider its basic nature, its source, the form it takes, and how it is generated. Threat types for space systems include physical (kinetic), electronic, optical, and cyber threats. The process of identification of the threat type often reveals the system elements against which it is effective and thus helps locate potential targets within the system.

#### 3.2.1.1 Physical or Kinetic Threats

Physical or kinetic threats are threats that can impact the system with kinetic force to cause physical damage and can be directed at all system elements.

Kinetic force can be manifested in the form of a projectile (hostile action) or a natural phenomenon (earthquake, tsunami).

The ground segment is vulnerable to a wider range of physical threats than the space segment, including natural phenomena and conventional projectile weapons such as missiles, up to and including armed invasion.

Adverse conditions such as earthquakes can strike without notice, while severe weather may be forecast in advance, allowing mitigation actions to be taken. If a hostile action is successfully executed, the impact results in the loss of functionality of the ground element. Attacks can be launched from air, sea, or ground depending upon the location of the ground site and its specific vulnerabilities. An attack need not necessarily result in complete destruction or incapacitation of the entire facility but simply disable key functionality sufficient to cause service interruption. For example, targeting an antenna or power lines could be sufficient to cause loss of service of a ground site.

Space segment threats were once thought to be rare and largely restricted to natural phenomena such as meteoroid or debris impacts. More recent threat analyses now include human produced threats including direct ascent ASAT missiles launched from Earth and co-orbital ASATs that could attack from space.

Physical threats, as with other types of threats, need not result in permanent, irreversible damage or disruption to the system element targeted. Causing a capability loss during a critical time period could result in enough disruption to be considered successful.

### 3.2.1.2 Electronic Threats

Electronic threats are those in which electromagnetic (EM) transmissions at radio frequencies adversely affect the performance of electronic and RF components. These electronic threats could result from hostile actions (intentional jamming), unintentional actions (accidental interference), or even be naturally generated (certain types of noise). Intentional jamming at lower power levels is intended to disrupt operations, while high power jammers are designed to damage sensitive receiver hardware. Threats are most often directed at satellite uplinks but can also be directed at ground sites and even user terminals. Terrestrial microwave links can also be the target of such interference.

Electronic jammers must be tuned to the same operating frequency as is used by the space system in order to obtain their greatest advantage. Higher power levels on a particular frequency result in more severe effects to the signal channel performance. A given jammer has a fixed amount of power to devote to its mission and the strategy for using that power most effectively varies by the target. The simplest jammers concentrate their energy at a single frequency, a waveform known as *continuous wave* (CW). This provides a high average power, but most efficiently interferes with narrowband signals;

wideband signals can be less affected. CW jammers can also be employed to saturate the satellite receiver, overwhelming and suppressing the desired signals. An alternate strategy is to spread the power over a wider band of frequencies to better interfere with multiple signals in the band and/or wider bandwidth signals.

The theory and practice of communications jamming and countermeasures is a broad topic of its own [8]. Many types of jammers are designed to deny service to SATCOM systems. These include full band, partial band, pulsed, and frequency hopping jammers. Sophisticated adversaries with significant resources can also deploy multiple jammers operating in the same frequency band and use them against a single target to increase the severity of the threat through cumulative signal power.

In addition to interfering with satellite data links, satellite command uplinks are also potential targets. Electronic threats can produce denial of satellite command and control by jamming uplink frequencies, the transmission of unauthorized commands, and interception of telemetry can provide information regarding the state of the satellite, potentially exposing vulnerabilities. The latter threat, monitoring of satellite data transmissions, is often referred to as *traffic analysis* and is used to determine operational information through indirect evidence via electronic observation of the data flows. Modern military satellites have provisions to defeat these threats, including encryption of telemetry and command data, however, protection varies for commercial satellites.

### 3.2.1.3 Optical Threats

Optical threats employ electromagnetic waves at optical wavelengths, often through the use of high-powered lasers. Lasers develop highly directed signals using coherent light sources and thus can be used to precisely target a point at great distances without losing significant power due to divergence of the beam. These threats are exclusively due to malicious activities and affect optical sensors and, if severe enough, can result in physical damage as well. As with electronic threats, optical sources must be matched to the operating wavelength of the space system to be effective when aimed to interfere with sensor operation.

Theoretically this type of jamming can be highly effective because of the high directional gain of an optical system, enhancing the effects of an optical jammer. In practice, the technique can be difficult to execute successfully for the same reason. Optical payloads have a very narrow field of view due to the use of coherent beams of light and thus the pointing must be very precise. As a result, it can be very difficult to place an optical jammer within view of the optical payload to obtain the desired effects.

High energy optical threats, also known as *directed energy threats*, cause degradation through thermal exchange to the satellite's radiators or solar cells, or by causing physical damage to other vulnerable parts of the satellite.

These energy weapons can be operated at wavelengths that cause the maximum damage to its target and do not depend on being within the field of view of an optical aperture.

### 3.2.1.4 Cyber Threats

Cyber threats are wholly the result of hostile actions and affect information systems and computing networks. These threats take on many forms and the attackers may have any number of goals in penetrating the network. Examples of (terrestrial) cyber attacks are many and include distributed denial of service (DDoS) attacks, distribution and propagation of computer viruses, Trojan horses, ransomware, data manipulation, data destruction, "person in the middle attacks," installation of spyware and malware, and advanced persistent threats (APTs). The aggregate number of entry points into the network that represent targets and must be secured or monitored is often referred to collectively as the *attack surface*, and various strategies exist as how to best protect the system at this boundary.

In contrast to other threats, cyber threats can generate a very wide range of effects. While the most common concern is the disabling of a network resulting in loss of command and control, many other actions may be taken once a network is penetrated. This includes the loss of sensitive data, the falsification of data, and the eavesdropping of ongoing communications across the network. In each of these cases the actions can occur without the knowledge of the network operators with even greater consequences as a more obvious attack.

There are many sources of and reasons for cyber threats. Commercial operators may be the target of extortion while government systems may be targeted for political and military reasons. Given the increasing reliance of space systems upon computing elements and networking (particularly in the ground segment), the risk of cyber attacks is considered to be increasing globally for all users. In addition, as the space layer becomes more of an extension of terrestrial networks, the existing vulnerabilities could find their way to satellites unless precautions are taken as the network functions migrate to space.

### 3.2.2 Threat Severity

A successfully implemented threat's impact on the system capability depends in part on the severity (or level) of the threat. The threat severity is based on the capabilities of its source and represents the extent to which the threat can cause service or capability disruption or denial. In the case of an adversary, there are many potential constituents of this overall assessment. Often the threat severity is linked to the technical sophistication (and sometimes economic might) of the adversary.

A near-peer adversary likely has greater technical and economic resources required to mount a more severe attack than an independent rogue agent.

For hostile actions it is important to consider all sources of attack, not just the worst case. Though a lesser adversary might not be capable of inflicting the same amount of damage as a greater one, its impact may still be significant. Any number of actors can have the motivation, means, and opportunity to cause some level of disruption, not only near-peer adversaries. In addition, the likelihood of a less severe threat being realized could be much higher than that of a more severe threat. Paying attention only to the worst case based solely on threat severity will not assure the desired resilience across the threat range given that lesser threats are credible.

Severity can impact the effectiveness of all of the four resilience attributes when matched against a system, but the most likely impacts are upon avoidance and robustness. The expected residual capability level (and thus the resilience) often decreases as the threat severity increases, as shown in Figure 3.4. At some point the threat could cause enough reduction in capability that the minimum mission requirement for resilience (and performance) is no longer met.

Threat severity can be further expressed in terms of intensity, persistence, and prevalence. The *intensity* can be described by the damage or disruption level that an attack provides—for example, the transmit power of an RF jammer aimed at a satellite. *Persistence* is the duration over which the attack occurs. A kinetic threat to a satellite is an example of an abrupt event, though the effect of the impact lingers. The use of a jammer over a long period of time is an example of a more persistent threat. *Prevalence* is a measure of the proliferation of the threat—the number of independent threat sources and/or targets. In a multiple jammer example, both the transmit power of each jammer and the number of jammers both contribute to the severity of the attack.

In this book, threat severity is often expressed as a *threat level*, which describes the number of system elements that are targeted by an adversary,

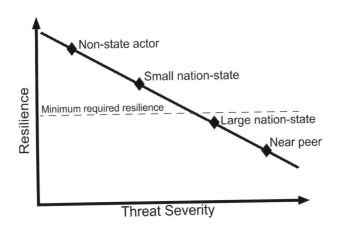

**FIGURE 3.4**
Resilience decreases with increasing threat severity.

either singly or in some coordinated fashion. The threat level is often the independent variable, used for parametric analysis. Of interest is how the system's resilience is affected across a range of threat severity.

### 3.2.3 Threat Target(s)

Each independent threat to a space system can be directed at a satellite, ground station, network connection, mission planning element, user, or all of these. The system response to each threat varies, depending upon the resilience of the affected system element. Figure 3.5 shows an example of a space system with two identified threats: the first to a satellite, and the second to a ground gateway. In this simple case, no single threat impacts multiple element types. In some systems, the system response to a threat is limited to the affected element. For example, an attack on a satellite results in actions taken at the satellite level in response. In other systems, the response might involve more than one element or even more than one system. This is important when calculating the mission-level resilience, as system recovery could involve multiple elements, or even other enterprise systems.

Note that in this example the threat mitigations against each of the two threats vary according to the targeted element. Steps taken to safeguard a satellite against a particular threat are likely to be quite different than those safeguarding a gateway against a terrestrial threat.

As long as the threats are independent, the analysis simplifies due to the fact that the local effects of each threat response can be superimposed to create the system-level capability response versus time. Each threat results in its

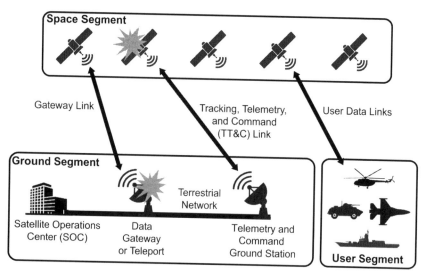

**FIGURE 3.5**
Example of targeted threats.

own impact, all of which can be added to calculate the total lost capability. In many cases, with a single threat, the local impact is the system impact as well, and the responses are the same. This is illustrated in later examples.

### 3.2.4 Threat Effectiveness

Another threat parameter is its effectiveness. Effectiveness is the probability of the threat being successful once it is activated (in the absence of mitigation). An ideal threat, when activated, would yield the desired result 100 percent of the time, but not all threats are 100 percent effective. Any number of factors can influence how successful a threat will be when exercised, not the least of which is the reliability of the system delivering the threat. An estimation of probabilities associated with the success of a given threat is important in creating a true picture. What are inhibitors to the threat being initiated? How reliable is the system delivering the threat?

In practice the effectiveness of real-world threats is often limited by any number of factors beyond the control of the adversary. If a rifle is solidly mounted to a fixture and aimed at a one-foot square pane of glass from a distance of six feet inside a closed room, the probability that when the trigger is pulled the bullet will strike and penetrate the glass is very nearly 1. There is a small chance that the rifle might malfunction and not fire when the trigger is pulled. If that same rifle is aimed by a person at the same pane of glass outdoors at a distance of 1000 feet, the probability of striking the glass is lower. The wind might affect the path of the bullet. Gravity significantly alters the trajectory over the longer distance. The accuracy required is greater and more dependent upon the skill of the person aiming. The effectiveness of the threat, despite the fact that no mitigations have been employed, is less than for the first case.

Once a threat is initiated, the worst-case scenario for the system operator is that the probability that it achieves the desired result is 1, meaning that it is completely effective. In this regard, the perfect threat is a special case and tends to reduce the complexity of assessing system resilience. An example is making the assumption that once targeted, a satellite or ground station is completely disabled, with no opportunity for recovery or reconstitution. Again, the threat effectiveness is separate from any threat mitigation features added to increase resilience. It is the probability of a successful attack in the absence of any countermeasures.

In military terminology, this parameter is called the *probability of kill* ($P_k$). It is important to recognize that in practice $P_k$ is likely to be less than 1 and using the worst case could cause the system to be overdesigned. A threat assessment might determine that there are flaws in an attack mechanism which limit its probability of being effective, particularly in certain scenarios or environments. Or it could represent uncertainty of success based on environmental factors outside of the control of the adversary. As a result, it is important to maintain this parameter as a variable, and to observe how it affects the system design in terms of the resilience obtained as a function of the threat.

Admittedly, gaining enough insight into the nature of a space system threat to accurately assess its effectiveness is difficult. Often this type of information is obtained through multiple observations of one or more events leading to statistics that enable an estimate to be made. In the absence of a statistically significant sample space, the effectiveness might have to be inferred through other means and technical intelligence. Suffice it to say that the error bars on this parameter are likely to be wider than for others used in resilience calculations.

For adverse conditions, observations also form the basis for establishing statistics that lead to an estimated effectiveness of the threat. The mean size and prevalence of orbital debris in certain orbits could be used to gauge both the severity and the effectiveness of a physical co-orbital threat.

### 3.2.5 Persistent Threats

In some scenarios a threat is considered repetitive or *persistent* over some period of time. In this case, assuming the system has a reasonably high level of resilience, the interest is in how long the system can continue to maintain its minimum required resilience in the presence of such a threat, particularly if long-term exposure to the threat can result in cumulative damage or degradation over time. Sometimes this duration parameter is referred to as *survivability* (not to be confused with *nuclear survivability*). The longer the minimum capability is retained, the greater its survivability.

If the system's capability does not in any way deteriorate or degrade as it sustains the persistent threat, then the survivability could be as long as the mission life of the system. However, if the persistent attack causes a gradual degradation of the capability over time due to exposure to the threat environment, then the duration over which sufficient mission capability can be maintained represents a quantitative measure of the system's survivability to the associated threat. This results in a threat assessment of resilience versus time. The natural radiation environment in space is an example of a persistent threat with effects that accumulate over time to cause damage if not mitigated through shielding or other means.

### 3.2.6 Reversible and Irreversible Effects

A threat's impact may or may not result in permanent and irrecoverable damage to the system. Threats that do not cause permanent damage produce *reversible* effects. That is, once the threat event has ended, the system (eventually) returns to its nominal operating state without serious degradation. For example, RF jammers can produce enough power to completely interrupt service, but yet not enough to damage the satellite's receiver.

Some threats, when successfully executed, cause permanent damage to the system, and these are *irreversible* effects. Examples include nuclear explosions, orbital debris collisions, and some extreme space weather. Following such events, the system can be partially or fully disabled. Some or all of the lost capability might be recovered either through autonomous operations or through human operator intervention, with the remaining capability irreversibly lost.

Certain threats can result in irreversible effects that do not cause permanent damage to *hardware*. Cyber attacks can result in data loss and damage to software housed in computing systems which might not be recoverable. If the data was not backed up and is permanently lost, the effect would be considered irreversible.

The mitigation approaches for reversible and irreversible effects differ and thus it is important to understand the available recovery and reconstitution functions going forward.

## 3.3 Multiple Threats to a System

The resilience of a system is likely to vary depending on the specific threat and most systems are designed to sustain multiple threats. As a result, the process of evaluating a system's resilience involves evaluating the impact of each threat and capturing each resulting resilience value. The result is a compendium of resilience values by threat. Suggested numerical methods for dealing with multiple threats is provided in Chapter 5, depending on the nature of the threat scenario.

While it can be desirable and convenient to collapse the entire resilience metric into a single number, that is not necessarily practical or possible. The resilience evaluation can span multiple values, one for each threat scenario, and the collection of values likely provide a better picture of the system resilience than a single value, which can mask certain useful information.

## 3.4 Evolving or Escalating Threats

It is tempting to consider threats as constant and stationary based on the best available knowledge during the system design phase. But today's human-generated threats are based on and in many cases limited only by the state of the art of relevant technology. As this technology inexorably advances, the threats also escalate. Projection of future threats and the evolution of the existing threats is a valuable exercise due to the relatively long lifetimes of space systems, typically 10 years or more.

While it is true that forecasting carries its own risks and inaccuracies, the act of considering threat evolution can result in devising architectures and/

or adding mitigation features such that future system evolution can more easily and rapidly compensate for the threat level. One measure of a system's ability to sustain escalating threats is a term known as *fragility,* which is described by the amount of excess resilience that the system has against a threat. Smaller margins increase the fragility of the system. If this margin is high, the system could be resilient enough to sustain some amount of increased threat severity.

Systems that are more modular, flexible, agile, scalable, reprogrammable, and reconfigurable are more likely to be capable of adjusting to escalating threat levels. These attributes make it possible for the system designers to keep up with new and evolving threats through a more continuous, incremental system upgrade cycle rather than having to contend with longer upgrade cycles with more extensive changes that could open vulnerability gaps. Figure 3.6 illustrates the comparison between these two types of long-term system development approaches. While both the shorter and longer refresh cycles ultimately provide similar system resilience at the end of the time period, the shorter refresh cycle has enabled the system's resilience to outpace the threat for a greater percentage of time than for the longer refresh cycle over this period. Any additional cost of supporting a greater number of shorter cycles must be balanced against the value of higher resilience over this time period.

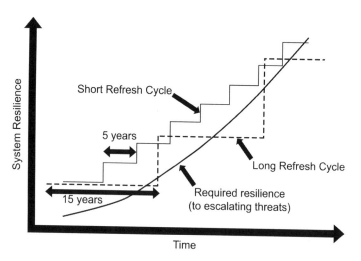

**FIGURE 3.6**
System resilience must keep pace with escalating threats over the mission life.

# 4

## *Threat Mitigation*

Once the system requirements, including the required resilience to specified threats, have been defined the designer can begin the process of generating implementations and executing trade studies to evaluate the performance of a range of candidate solutions. In addressing the required system-level resilience, the designer must consider the available mitigation options that are effective in countering the projected effects of the threats on the system. This becomes the designer's tool box from which to draw to increase system resilience. The resilience values that each of these mitigation features provides are estimated and used in the calculations of system resilience, as discussed in Chapters 5 and 6.

System threat mitigation requires a detailed analysis of each threat and determination of the potential for disruption to operations. In response, the designer must choose and incorporate design features that counter the threats by mitigating their intended effects. There are many ways for the system designer to choose and incorporate design features that counter the threats, both at the element level and through the choice of system architecture. In addition to specific features that protect against specific threats, there are also certain system qualities that provide resilience. These include, but are not limited to:

- **Flexibility:** The extent to which the system can be reconfigured or operated in different ways to compensate for the effects of the threat
- **Agility:** How quickly the system's state can be reconfigured in response to a threat to mitigate its effects
- **Scalability:** The ease with which a system architecture can be scaled in response to an evolving or escalating threat
- **Modularity:** The ease with which system elements and components can be replaced or modified to adjust the system's behavior to meet new or changing threats. An example is the U.S. DoD Modular Open Systems Approach (MOSA) initiative, which encourages the use of open standards and interfaces to enhance system modularity.
- **Extensibility:** The extent to which new capabilities or features can be added to enhance the performance and/or resilience of a system when new requirements emerge, particularly in a short period of time

This chapter addresses the opportunities for threat mitigation and discusses some of the mitigation approaches for several types of threats that are most likely to be encountered for space systems. For large, complex systems with many space and ground elements multiple mitigation features might be required to provide the required resilience. Likewise, multiple threats of different types usually require a combination of different mitigation features to ensure that a high level of resilience is achieved, as many mitigation features are specific to each type of threat.

## 4.1 Threat Mitigation Approaches

Threat mitigation is predicated upon evaluating the credible threats to the system and identifying likely targets and vulnerabilities that must be reinforced. More discussion of the detailed process is provided in Chapter 6. Specific threat mitigation opportunities are available at different points in the space system's operational timeline. A representative behavioral timeline was presented in Chapter 2 showing the response of the system from steady-state operations to an activated threat. Similarly, the threat mitigation approaches can be mapped against a similar operational timeline as shown in Figure 4.1. This timeline consists of two periods defined by the *event* time. An event is defined as the impact on the system due to a realized threat when the threat has not been avoided. For example, if a direct ascent ASAT is launched at a satellite, the event would be the collision of the weapon with its target. The period prior to the time of the event is the *pre-event phase;*

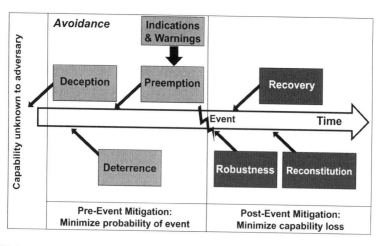

**FIGURE 4.1**
Timeline of threat mitigations.

the period after the event is the *post-event phase*. Different mitigation approaches are available for each of these phases.

*Pre-event mitigation* provides the earliest opportunity to counter threats using avoidance methods. The first method to consider is **deception**, which is specific to hostile actions and if successful provides avoidance by concealing the existence of the system capability from an adversary. The premise is that a capability unknown to an adversary cannot be a target. However, in practice it can be difficult to conceal the existence of a system while it is being actively used to deliver a capability, even if that capability is passive in nature (such as collecting overhead imagery).

Deception is considered first because if it is successful no other threat mitigation is necessary and ideal resilience is achieved. A drawback is the potential for the capability to be discovered at which time the system may be vulnerable to an attack if no other threat mitigation features are implemented. Stealth technology used to reduce the radar profile of fighter jets is an airborne example of deception employed to military advantage.

Avoidance can next be achieved through **deterrence**. Deterrence can be viewed as eliminating an adversary's *motivation*. The goal is to make the apparent cost to an adversary so high relative to the expected benefits as to deter an attack upon the system [9]. Quixotically, one of the primary means of achieving deterrence is through the deployment of a highly resilient system with threat mitigation features that ensure that any attack mounted would be fruitless. Thus, threat mitigation must still be implemented to ensure deterrence is effective.

The word "apparent" is used to stress that the external representation is what is important, not necessarily the actual implementation of the threat mitigation capabilities nor the actual resilience of the system. Perception is reality. Deterrence is a difficult parameter to assign a numerical value to, as the evidence necessary to make such a judgment is difficult to obtain and can be based on doctrine as opposed to data. This is not normally a resilience attribute that the designer has access to but is presented for completeness.

If neither deception nor deterrence is successful or practical, the next opportunity for threat mitigation in the timeline is **preemption**. Preemption is the act of removing an adversary's *opportunity* to carry out a threat. Preemption is accomplished by acting prior to the threat event (and possibly even prior to the initiation of action) to ensure that the space system is not impacted and is another means of achieving avoidance and providing protection. An example is a preemptive strike on an ASAT missile launch site prior to the launch taking place. Note that this approach requires information relating to the source of the threat, such as location, direction, timing, and other related data and also has political and military consequences. This data could be provided by a separate detection system providing indications and warnings (I&W). If resilience is achieved through this method then there is also the resilience of the I & W system against potential threats to consider, adding complication to the resilience evaluation.

Preemption can occur at the source (on the ground, in the previous example) or anywhere prior to the point at which the system behavior is changed, all the way to the destination. Satellite protection provided through the use of on-board protection (as defined in the OSD SDMA white paper [6]) is another method of preemption, but in space. All of these protection measures could include various protective actions and also require some knowledge of the immediate threat environment. These protection features can result in avoidance of the threat if successfully deployed. Note that in the OSD lexicon certain other types of countermeasures are excluded from the definition of resilience but can be equally effective.

Only threat mitigation measures prior to the realization of the threat are included in the pre-event phase. The threat mitigation might not be completely effective, in which case the probability of avoidance is some value less than 1. The development of mitigation features also does not consider the inherent effectiveness of the threat itself ($P_k$), which is independent of the threat mitigation. Successful threat mitigation results in an adjusted threat effectiveness (adjusted $P_k$) reducing the probability of an event, ideally to zero.

If the threat is not avoided, then an event occurs and the post-event phase begins with the associated post-event mitigations available. The first of these is **robustness**, which includes the system features that minimize the loss of capability following the event. The most commonly cited examples are protection features that impart a high damage threshold for sensitive components, such as radio receivers. Likewise, ground installations can include significant structural reinforcements to minimize the effects of any physical attack. If the protection is sufficient, the specific design features allow the system to sustain the impact of a threat with little or no loss of capability. However, when considering robustness at the system level, there are also architectural options that can be used to impart robustness. Note that protection is *not* the same as robustness but can provide robustness; there are multiple resilience sub-elements that can provide robustness, not just protection. These include distribution, proliferation, diversification, and disaggregation. Balancing the distribution of capability and the protection of individual elements to provide resilience through robustness is a fundamental system-level trade that is discussed in detail in Chapter 6.

If the system is not sufficiently robust to meet the system resilience requirements, then capability is lost and the next mitigation method to consider is **recovery**. Recovery features enable the system to regain, either through human intervention or autonomously, some portion of the initial capability potentially restoring one or more missions. An example is switching to a redundant satellite payload unit in the event that the primary unit is damaged. The time between initial loss of capability and the conclusion of the recovery operation is referred to as the *recovery time*. Sometimes this term is used interchangeably with *outage time*, but the latter tends to imply that full capability has been restored, while recovery time does not imply that the "outage" has ended, only that it might be less severe if the recovery action was at least partially successful.

Finally, **reconstitution** can be employed to recover full functionality, usually by replacing a major system element, such as a satellite or ground station. The distinction in the DoD definition is that recovery provides restoral of some of the missions or services supported by the system, whereas reconstitution restores all. As with recovery, the time required to reconstitute is an important parameter when considering its impact upon system resilience.

Multiple threat mitigation features can be combined to provide the required system-level resilience, as individual features might be insufficient. Mitigation can be achieved in any number of ways, each proving to be equally effective, though the cost and impact to system performance will vary. This timeline does not imply a priority order in which a designer should consider implementing the various mitigations. Instead the entire timeline should be considered with all mitigations available either singly or in combination to meet resilience requirements.

Note that none of the resilience attributes are intrinsically "better" than the others. Ease of implementation, cost, and ubiquity of value across multiple threats can all drive the decision as to which should be used to increase resilience. A key goal should be to obtain the maximum level of resilience against all threats at the lowest cost while maintaining the highest performance over the life of the system.

## 4.2 Threat Mitigation Options for Space Systems

As threats to space systems continue to evolve, escalate, and proliferate it is both impractical and undesirable to attempt to provide a complete catalog of all possible threats that a designer might be presented with. However, several general classes of threats are either being encountered today or have been demonstrated as a real-world capability, as detailed in Chapter 3. The following is a summary of demonstrated threat mitigation techniques for the more common threat types encountered in space system design.

### 4.2.1 Mitigating Electronic Threats: Radio Frequency (RF) Signal Interference and Jamming

The space and ground elements of a space system are connected via RF signals. When a satellite (or ground station) receives undesired RF energy in a signal band that is being used by the system to transmit and receive data, this is referred to as signal *interference.*

Interference is a common and sometimes persistent threat for many space systems, particularly SATCOM systems. Intentional interference is due to an adversary's deliberate actions and is often referred to as *jamming.* Interference is more commonly unintentional and due to benign activities, including operator error, misalignment, equipment malfunction, and even blue force sources, yet can have the same effect of causing degraded performance and/ or system outages.

The greatest vulnerability is to satellite uplinks that receive signals from the ground, as geometry and other issues make deliberate jamming of down-links much less likely and effective. Users most affected are so-called *disad-vantaged users,* who are often operating terminals with less transmit power and smaller antennas and are thus more susceptible to being adversely affected by this interference.

Electronic interference usually affects two key communications capabili-ties: If severe, it can completely deny access to a communications link alto-gether, reducing system capacity. In less severe cases, the link might still be usable but the maximum data rate for specific users will be reduced. Frequently the ability to operate in the presence of interference is referred to as robustness and has similar meaning as with resilience.

A number of techniques can be employed to reduce the sensitivity of space systems to interference. The measure of communications performance in the presence of interference is the ratio of the received jammer power (J) to the desired signal power (S) that can be sustained, called *J/S* and expressed in decibels (dB). In addition, a second metric is the *standoff distance,* which is the minimum distance between the interfering source and a user that pro-vides acceptable performance. The goal is to be able to sustain the highest value of J/S at the shortest standoff distance.

In the case of hostile actions, an adversary can tailor the interference signal characteristics in an attempt to provide maximum disruption to the commu-nications link. Jammer signal types include full band, partial band, continu-ous wave (CW), pulsed, hop-along, and repetitive. In some cases, multiple mitigations can be required to defeat multiple types of interfering signals. Again, the greater knowledge of the nature of the interference, the greater the opportunity to neutralize its effects on system performance.

Different types of RF links can suffer different impacts due to jamming. As noted in Chapter 1, satellite command and control (C2) links are very impor-tant to enable the operator to maintain positive control of the spacecraft. Interference to these links can limit or fully deny service to users. These links generally operate at lower data rates and thus occupy less bandwidth. These are often called *narrowband links.* The narrower the bandwidth, the easier it is to interfere with the signal, so these links are particularly vulner-able. In addition, the satellite antennas are not highly directional with most using omnidirectional antennas that do not provide any spatial rejection of incoming interfering signals.

Higher rate user data is conveyed using wider bandwidths over *wideband links.* These signals are more robust to interference, particularly certain types of jamming, such as CW signals. Intentional jammers must spread their fixed amount of power across a wider band, decreasing its effects. Satellite antennas supporting wideband links are also directional, providing spatial isolation that can reject signals not near the ground station, user, or gateway. This also helps overcome unwanted RF energy. As a result, more mitigation options are available for wideband links.

There are a number of mitigation techniques that can provide resilience against RF interference. Most of these involve adding specific features to the design of the space system, particularly on the satellite. The system capability, threat(s), and the concept of operations (CONOPS) can also provide some measure of mitigation as well. The more common approaches include employing advanced antennas, signal waveforms, digital signal processing, and RF filtering to achieve the desired performance in contested and congested environments.

### 4.2.1.1 Spatial Isolation

Satellite antennas provide coverage to certain geographic locations within their field of view. Different antennas provide different coverage patterns, or beams. Some cover the entire visible Earth (Earth coverage), some provide more regional coverage, and some cover only narrow areas (spot beams) (Figure 4.2). Signals within the coverage area are magnified, while those outside of the coverage area are rejected (attenuated). The narrower the coverage area, the more amplification of the signal. The magnification difference between the inside of the coverage area (in beam) and the outside (out of beam) can be considerable depending upon the size of the beam and its specific pattern. This provides a way to geographically isolate the interfering

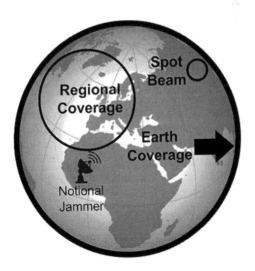

**FIGURE 4.2**
Comparison of satellite antenna coverage patterns.

signal from the desired signals. The narrower the beam, the greater the spatial isolation can be obtained for smaller standoff distances. The effect is to maintain good performance for the desired users by keeping the jammer outside of the antenna coverage area.

In the example of Figure 4.2, the notional jammer is within the field of view of the Earth coverage antenna pattern and could jam that uplink. It is located very near the edge of coverage for the regional coverage beam and can cause some degradation as a result of insufficient standoff distance, particularly for system users at the edge of coverage near the jammer. Finally, the spot beam would be unaffected, having significant robustness to the jammer due to considerable spatial isolation due to the smaller coverage area and greater distance from it.

The functionality of the satellite antenna can provide additional capabilities in mitigating interference. Some antennas are fixed in their position, covering a specific part of the Earth. Other narrow spot beam antennas can be steered, providing some flexibility in positioning the beam relative to desired and undesired signals. Certain electronically steered antennas, such as phased arrays, also allow complex shaping of the beams using ground commands. This enables more complicated patterns to be generated, providing higher performance to the users and shaping the beam away from a known interference source. In addition, both the received signal phase and amplitude can be adjusted for each of the many antenna receive elements, allowing signals from a certain geographic location to be rejected, or *nulled*. When a control loop and nulling algorithm is applied, this process can be automated and is called *adaptive nulling*. This is a proven method that can be used for systems employing phased arrays and multi-beam antennas (MBAs) as well. Though effective, this is a more costly method of achieving anti-jam performance.

Optical (laser) communications links represent an extreme example of spatial isolation. Communications signals can be carried by laser beams of coherent light. These beams are very tightly focused and diverge very little even over long distances. As a result, the beam width is vanishingly small and almost impossible to detect, intercept, or jam. The applications are more limited than for RF systems, as weather and other atmospheric effects can degrade the signal and reduce link availability, but selected use can significantly improve resilience to a jamming threat. In this case, the threat is more likely to be optical as well, but the spatial isolation makes this threat very difficult to successfully execute.

### 4.2.1.2 Receive Frequency Selectivity (Filtering)

Another common and cost-effective method of discriminating between desired and undesired received signals is through frequency selectivity. If a jamming signal appears in the receive antenna coverage area and cannot be otherwise spatially isolated, then frequency discrimination is another alternative. SATCOM systems operate in specific frequency bands (as shown in Figure 1.6), and the transmitters and receivers are tuned to those bands, rejecting signals outside of them. If the interferer is within the receive band, it is called an *in-band jammer*. Otherwise, it is an *out-of-band jammer*. Clearly the in-band jammer is more difficult to mitigate since it occupies the intended receive bandwidth alongside the desired user signals. The desired signals can be separated from the interfering signals using narrow filters in the satellite receiver thus reducing performance degradation. Sometimes these narrow bandpass filters can be tuned across a frequency band, providing flexibility if the interference moves in frequency over time. Tunable notch filters can also be used to reject a very narrow frequency range.

A figure of merit is how close in frequency an interfering signal can be to the desired signal without significantly affecting performance. This depends upon many things, including the selectivity of the satellite's receive subsystem. The better the channelization and filtering, the less the effect of the interference. Increasing the satellite, ground station, and user terminal selectivity all provide some measure of resilience against interfering sources. This is usually accomplished through very narrow bandpass filters with high selectivity similar to that shown in Figure 4.3. In the figure, the desired signal occupies bandwidth around the center frequency, $f_c$, in the filter passband which passes signals with very little loss. At the edges of the passband, the signal loss, or attenuation, increases very rapidly as bandpass filters typically have very sharp cutoff profiles. Out-of-band signals are rejected, eliminating potential interference to the desired signals.

Filter selectivity is measured as a ratio of the passband bandwidth to its center frequency, a figure called $Q$. High-$Q$ filters can provide high selectivity but sometimes at the expense of in-band signal loss and phase distortion, including increased delay. A more expensive solution is to employ on-board digital processing to provide increased performance through better channelization and more frequency mapping flexibility. Digital filters can

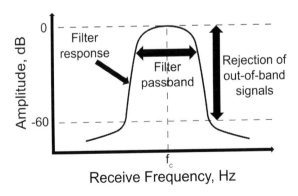

**FIGURE 4.3**
Narrow bandpass filter provides frequency selectivity and rejects out of band interference.

be implemented with near-ideal frequency responses but require processing power to do so. Nevertheless, this approach is gaining favor as the power efficiency of digital integrated circuits continues to increase.

### 4.2.1.3 Spread Spectrum Waveforms and Digital Signal Processing

The choice of signaling is also an important technique to increase resilience against interference. Different signal waveforms provide different levels of protection depending upon the type of interference. A method that spreads the user's signal power across a wide frequency band is known as spread spectrum signaling. There are multiple ways to accomplish this, including the use of direct sequence spread spectrum (DSSS) and frequency-hopped spread spectrum (FHSS) waveforms. In both cases, the goal is to transmit the signal in such a way as to make it highly robust to various types of interfering signals.

Figure 4.4 illustrates how a user's signal frequency is randomly hopped across a certain bandwidth to uniformly spread the signal power. Even though the user signal power is much lower than a jammer's power, the jammer only disrupts the user signal for a small fraction of time, and many systems use advanced error correction coding techniques to recover even that lost data.

FHSS systems are more complicated than traditional single frequency systems because both the transmitter and the receiver must be precisely synchronized in time so as to hop according to a random code that defines the sequence of frequencies that each must use concurrently. Cryptographic devices are required to provide these secure codes. Spread spectrum waveforms can provide another type of robustness. By spreading their energy over a wide bandwidth, detection of the signal becomes more difficult as the peak signal power approaches the channel noise level. This class of waveform exhibits a low probability of detection (LPD) and a low probability of intercept

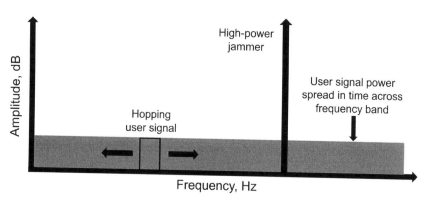

**FIGURE 4.4**
Frequency-hopped spread spectrum mitigates jamming.

(LPI). LPD waveforms are less susceptible to jamming, resulting in a measure of avoidance that also reduces the likelihood that it can be monitored.

Often additional digital processing of the source data is added to provide even greater robustness to interference, including forward error correction (FEC) coding and bitwise interleaving. These techniques reduce the error rate of the received data stream by enabling errors to be detected and corrected by the receiver. Again, each of these features comes with a cost, but the continuing reduction in cost of digital memory and processing power is lowering the cost of implementing these functions, both on the ground and in space.

### 4.2.2 Mitigating Physical (or Kinetic) Threats

Physical threats have most often been considered against ground infrastructure and user segments for space systems. However, more recent events demonstrating ASAT technology have raised concerns about physical attacks on satellites in orbit as well. Any system element can now be considered as a candidate for physical damage or destruction.

Mitigations against physical threats generally center on increasing physical security around the perimeter of the system element. For ground sites this can include walls and fences, reinforced structures with locks, and other common means of securing a facility, including human factors such as armed guards and electronic surveillance. For satellites, the options are somewhat different. Satellites are relatively fragile and physically reinforcing them is not practical for larger satellites due to the added mass penalty that would be involved. Robustness in the satellite context usually refers to adding metallic shielding to protect against radiation and nuclear effects rather than actual structural strength. Antennas and solar panels can be vulnerable to sudden motion, let alone a ballistic threat. Each system element possesses different vulnerabilities and characteristics and should be considered separately.

#### 4.2.2.1 Ground Stations and Terrestrial Networks

Most space systems rely on a global ground infrastructure to provide command, control, and connectivity between the satellites and the users. These ground sites, often known as gateways, are connected to terrestrial networks via coaxial cables or optical fiber. Both the ground sites and the terrestrial network are targets for disruption. Loss of either can result in a loss of system capability. Mitigation approaches for these types of installations can include, but are not limited to, the following:

- **Location:** Mitigations for ground sites start with choosing secure locations that offer protection against a variety of threats. The goal is to obtain both avoidance and robustness. Critical ground sites are located away from severe weather and natural hazards, such as active earthquake fault lines. The U.S. government often locates its ground stations on existing U.S. or allied military bases as one strategy to protect and deter. Ground stations are often heavily fortified to provide robustness, however, the need for large external antennas can introduce vulnerabilities. Both government and commercial systems feature backup sites for many of their ground components (especially for command and control functions) in the event of either an adverse condition (earthquake, storm) or a hostile action.

- **Mobility:** Mobility is another means of resilience applicable to ground elements. Mobile or transportable equipment can complicate adversarial targeting in an affordable manner. Some military systems use mobile command post terminals to improve resilience to the nuclear threat. Relocation of critical assets during a severe weather event, such as a hurricane or tsunami, is another example in which transportability can be an effective mitigation technique.

- **Redundancy/Diversity:** Networked redundant ground stations are another method of providing resilience. Many satellites are within the field of view of multiple ground stations. Though more expensive, geographic diversity can provide both deterrence and robustness if key functions can be replicated and executed from multiple locations distant from one another. As high bandwidth networking of ground stations has become more affordable and available, the options for adding agility to operations have increased, making the transition of services from one ground site to another much less labor intensive and more responsive. Network connections are also primarily made resilient through the use of redundant paths. Use of multiple parallel fiber or coaxial cable lines between sites limits the impact of the loss of a line whether through failure or hostile or benign action.

- **Inter-satellite crosslinks:** Some satellite systems, particularly the strategic ones, have eliminated the need for multiple ground stations altogether, instead using RF crosslinks between satellites to provide global connectivity. The U.S. Milstar and Advanced EHF (AEHF) strategic MILSATCOM systems take this approach, as does the commercial Iridium system. Spacecraft command and control (through the AFSCN and other sites) is still required, but the satellites can operate autonomously for a period of time without any ground contact at all, adding resilience to the system. The 66-satellite Iridium global communications system uses RF crosslinks between satellites to relay data and reduce the number of required ground stations. This crosslink approach is effective because these satellites use on-board processed payloads capable of performing many key switching and routing functions often found on the ground. Though this approach increases the cost of each satellite, the benefit is reduced global ground segment operations and maintenance costs.

### 4.2.2.2 Mission Planning Element

A mission planning system function is required to schedule and coordinate the operations of the entire space system. The loss of the mission planning element can cause significant disruption to user services even though the rest of the system is unaffected. As a fixed ground site, it suffers similar vulnerabilities as a ground station, although as most government and commercial mission planning sites are within secure borders, the threat might not be considered to be as great as for those outside of a nation's own borders, with the possible exception of cyber threats.

Mitigation options for the physical attacks on the mission planning element of a system are similar as for other parts of the ground segment but can be more limited. This is because while multiple ground stations and command and control sites often exist there is usually only one or two centralized mission planning sites for a given space system. This makes the need for resilience of the mission planning element even more acute due to the possibility of this function being a single point failure for the system. It also suggests implementation of autonomy functions for the remaining system elements, particularly the satellites, to remove the dependence upon the mission planning element (MPE) for as long as possible. This could include such features as the ability to upload long-term schedules to on-board memory on a satellite to enable it to execute for some period of time in the absence of a ground contact.

### 4.2.2.3 Space Segment

Physically attacking satellites through the use of kinetic weapons is a relatively new threat and one that continues to evolve. While there is little direct public evidence to provide the effectiveness of this threat, there is still a concern because modern systems have not been designed to mitigate it. While many threat mitigation options against ASATs and other weapons in space have been suggested for over 30 years, actual development appears to still remain in the conceptual stage so the range of alternatives, while likely growing, is largely yet to be proven. Some of the more popular alternatives that have been suggested include the following:

- **Maneuverability:** The most promising techniques are passive ones, allowing the satellite to avoid an incoming threat. Today's GEO satellites are designed to be moved very slowly once on orbit, and likely not fast enough to evade a guided projectile approaching at high speed. Improving a satellite's maneuverability can provide additional resilience. However, this can require additional fuel, perhaps new types of propulsion, and added cost. Today's satellites are not designed to move quickly and might suffer damage even if they could. Nevertheless, there remains interest in this area.

- **On-board protection:** On-board mitigation features are also an option, however, the satellite must be made aware of the threat and have all of the information to activate a feature to deal with it. This approach relies upon some means of detecting and targeting an incoming threat and then actively defending against it.

- **Deterrence:** Policy and doctrine could be put in place to deter an adversary from pursuing an attack on a satellite. The consequences could be so harsh as to essentially eliminate the threat entirely. Alternatively, the satellite might be chosen to have material or economic value to the adversary to the degree that an attack is deterred—for example, use of a commercial satellite that also supports the adversary's critical economic infrastructure. Generally, this approach would be in addition to more conventional engineering solutions.

- **Location:** Use of nontraditional orbits, such as a *super synchronous orbit* farther away from the Earth than the GEO arc (22,236 miles altitude), can provide a certain level of protection and/or situational awareness leading to increased probability of avoidance. As with other mitigation approaches, there are performance trade-offs involved. For example, use of a super synchronous orbit increases the time delay for SATCOM systems and the latency of data relayed, which can be an issue when connecting into some networks.

- **Use of novel or unique orbits:** As discussed in Chapter 1, choice of satellite orbit can have an effect on the vulnerability to certain threats and impact resilience. Choosing orbits that provide some measure of protection or lower the probability of being targeted is yet another way to mitigate threats.

- **Use of covert assets:** The use of covert assets is another method of achieving at least limited avoidance. If the capability can be hidden until such time as it is required, then the probability of an attack is very low. However, this method requires investment in a resource that likely provides a one-time advantage, as the activation of the capability informs an adversary as to the existence of the capability, making it a target and limiting future utility.

### 4.2.3 Mitigating Optical Threats

Systems that rely on optical sensors are vulnerable to optical threats. Ground-based optical transmitters can generate high power optical beams, often using laser technology to tightly focus the energy at a target at long range. Key threat parameters are the optical wavelength (equivalent to frequency for RF), peak power, and average power (often related to the duty cycle, repetitive pulsed versus continuous wave).

The threat must be matched to the sensor's wavelength, because otherwise it is an out of band threat and only a small percentage of the incident optical power is absorbed by the sensor to cause an outage or damage. Lower transmit power levels can result in the temporary blinding of the sensor (sometimes referred to as "dazzling"), while very high power levels can cause irreversible damage, or burnout of the sensor.

As with electronic threats, the optical receiver is most vulnerable to the threat. Satellite optical receivers often consist of a semiconductor sensor with some type of focusing optical assembly to capture and focus the light, such as a telescope. Examples are the Hubble Space Telescope and the Space Based Surveillance System (SBSS). The field of view of the receiver can be wide or narrow. Again, as with RF systems, the wider the coverage area, the lower the amplification but the more susceptible the system is to an optical threat.

Mitigation techniques can include adjustable optical filtering to narrow the bandwidth seen by the sensor, reducing the received incident power. And some types of CONOPS in conjunction with a moveable blocking shutter mechanism and sensing of the threat prior to exposure to the incident beam have also been suggested.

### 4.2.4 Mitigating Cyber Threats

The network segment can be vulnerable to cyber threats. Many commercial and government terrestrial space networks are international in nature and susceptible to hostile actions. Network system administrators and cyber security specialists must constantly work to maintain system security. Failure to do so can result in compromised or lost data, loss of control over system elements (including satellites), and disruption of services to users.

Mitigation approaches are varied and often multilayered. First and foremost is the establishment of some measure of physical security for sensitive elements of the system. Many cyber attacks are predicated upon unrestricted access to hardware. Other mitigations include the use of firewalls, virtualization, and hypervisors and cognitive processes to detect intrusions and deny access. Identification of intrusions is of particular importance such that an appropriate mitigation can then be selected based on the activities of the intruder.

While the discipline of cyber resilience is consistent with the resilience concepts previously discussed, the implementation is singularly unique and the specifics are not contained within this book, as there are a number of other sources that provide a more detailed discussion of the topic [10].

# 5

## Modeling and Calculating Resilience

System designers require metrics that can be used in trade studies to evaluate their candidate designs and select the optimal solution. Resilience is one such metric. Some metrics result in very precise results, while others are less accurate due to the quality of the information used in the analysis. Resilience is no different, producing an estimate of the system behavior in response to a specific threat or scenario. This chapter presents a methodology based on simple mathematical models of a space system's behavior. This is ultimately used to derive a resilience value that represents the expected value of the residual mission capability for exposure to a specific threat.

Given the goal of making relative comparisons of the merits of different system designs, this metric is useful though it does not necessarily predict the exact amount of residual capability following an event, depending on the threat and the system design. Certain special cases are truly deterministic, with accuracy only limited by the available information and its precision. In other cases, the value will be probabilistic, an expression of an expected value. Many scenarios will represent special cases of the more general expression.

In addition, some parameters such as recovery are most useful when the value of time is considered. Given that recovery occurs after some outage time, designers can choose to place some manner of value weighting on it, adjusting the recovered capability parameter value by a function of the outage duration. Even if full capability is restored at some point in time, the recovery value might be less than 1 simply because of derating according to the length of this duration. Shorter outages are more desirable even if two systems eventually recover to the same capability level.

As a result of these practical considerations this approach to calculating resilience results in an *estimation metric, and not necessarily an absolute prediction* of the eventual final system capability following an event. If that value is desired, it can be found as part of the calculation process but might omit some of the criteria that have been levied against resilience in its definition.

## 5.1 Modeling Resilience

The DoD definition of resilience clearly identifies the amount of capability lost and the duration of that loss to be two important parameters. A very straightforward calculation integrates the time domain capability function, if known,

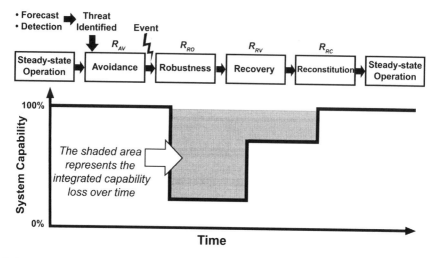

**FIGURE 5.1**

Simple integration approach to calculate a resilience value.

**Source: "A Method for Calculation of the Resilience of a Space System," MILCOM 2013, R. Burch, 2013.**

to find the area under the curve over some defined finite time period (Figure 5.1). This is a simple, yet acceptable method though it requires knowledge of the detailed system capability time domain system response to the threat, which might be difficult to obtain, and avoidance must still be handled separately.

A more extensible method that incorporates the avoidance attribute into the final resilience value has advantages over this approach. Using the definition of resilience provided in Chapter 2, it can be asserted that system resilience, R, should be in some way a function of the four attributes previously described. If each attribute is assigned as a variable of the resilience function R, then we have Equation (5.1):

$$R = f\left(R_{AV}, R_{RO}, R_{RV}, R_{RC}\right) \tag{5.1}$$

In the case of all variables except avoidance, each can be defined as a fractional measure of capability, either retained, recovered, or reconstituted. Thus, these three variables can take the range of values between 0 and 1, corresponding to an effectiveness between 0 and 100 percent for each. For example, a robustness value of $R_{RO} = 0.75$ means that when impacted by a specific threat event, 75 percent of the system's initial capability is retained. For recovery, 0.75 would represent a recovery of 75 percent of the lost capability.

Flowing the mathematics through the timeline shown in Figure 5.2, the algebraic Equation (5.2) can be created to represent the resilience value of a system, $R_i$, *for a threat that is not avoided*:

$$R_i = R_{RO} + \left(1 - R_{RO}\right)R_{RV} + \left(1 - R_{RO}\right)\left(1 - R_{RV}\right)R_{RC} \tag{5.2}$$

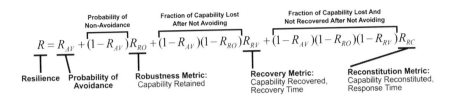

FIGURE 5.2

The resilience equation as a function of four variables.

Source: "A Method for Calculation of the Resilience of a Space System," MILCOM 2013, R. Burch, 2013.

Since each of the variables can take a value between 0 and 1, it can be shown that the range of the resilience value $R_i$ is between 0 and 1 as well. This value is the retained fractional capability following both recovery and reconstitution. If any of the three variables equals 1, $R_i$ also equals 1. But this equation does not yet include the avoidance parameter, $R_{AV}$.

Equation (5.2) would suffice if the scenario is constrained to allow the system to either avoid the threat, or not. But avoidance is one variable that is likely to have some uncertainty and thus it is desirable to represent it as a probability. If avoidance is probabilistic, it will also take on a value between 0 and 1, representing a *probability of avoidance*. When avoidance is also considered, the resilience value becomes an expected value of residual capability. The expected value $E(x)$ for a random variable $x$, is defined mathematically as:

$$E(x) = x_1 P(x_1) + x_2 P(x_2) + \ldots \tag{5.3}$$

over all discrete $n$ values, $x_n$, that $x$ can take, and where the sum of the probabilities $P(x_n)$ is 1. Here our random variable is the avoidance variable $R_{AV}$. There are only two cases to consider due to the definition of avoidance being to *fully* avoid (or not). If the threat is fully avoided (with probability $R_{AV}$) there is *no loss in capability*. In the case in which there is no avoidance (with probability $1 - R_{AV}$), the loss is the value found from Equation (5.2), and the term is that value multiplied by $1 - R_{AV}$. So, the resulting equation to find the expected resilience value for these two outcomes becomes:

$$R = E(x) = (1)(R_{AV}) + R_i (1 - R_{AV}) \tag{5.4}$$

The resulting expanded resilience equation is shown in Figure 5.2. The inputs are the aforementioned variable values that are derived from the system or architecture design and the knowledge of the threats.

This equation has two desirable mathematical properties:

1. Excluding the deterministic special cases where the avoidance value is 0 or 1, the result of the equation is probabilistic, making the resilience an expected value. This is a consistent interpretation because

in most cases uncertainties in the threat properties and/or system response should result in a probabilistic value for resilience. The avoidance parameter *can* be 0 or 1, although that is often a simplifying assumption.

2. The equation is symmetric in all four variables. This means, for example, that if any of the four variables equals 1, the system resilience is 1. This is important as it reinforces the fact that each resilience attribute is mathematically equally weighted (though the implementation costs will likely differ). This symmetry extends to the first partial derivatives of the resilience equation as well.

Alternate forms of the equation confirm this property: The symmetry of the equation in all four variables can be easily seen through the full algebraic expansion (using representative variables, as in Equation (5.5)):

$$R = \left(w + x + y + z\right) - \left(wx + wy + wz + xy + xz + yz\right)$$
$$+ \left(wxy + wxz + wyz + xyz\right) - wxyz \tag{5.5}$$

For greater ease of computation, the equation can also be written in factored form, as shown in Equation (5.6):

$$R = 1 - \left(1 - R_{AV}\right)\left(1 - R_{RO}\right)\left(1 - R_{RV}\right)\left(1 - R_{RC}\right) \tag{5.6}$$

The resilience Equation (5.2) can also be viewed as a sequential construction of terms. A process flow chart reflecting this view is provided as Figure 5.3.

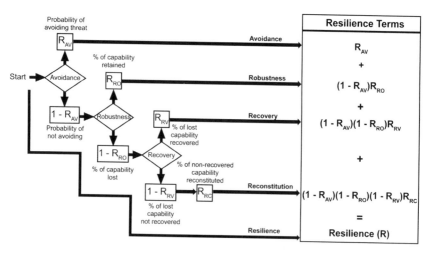

**FIGURE 5.3**
Resilience calculation flow chart.

| Segment | Threats | Avoidance | Robustness | Recovery | Reconstitution | Resilience |
|---------|---------|-----------|------------|----------|----------------|------------|
| Space | RF Interference | 0.100 | 0.250 | 0.800 | 0.000 | 0.865 |
| | Nuclear Event (in space) | 0.000 | 0.950 | 0.900 | 0.000 | 0.995 |
| Ground | Physical Attack on Ground Station | 0.850 | 0.900 | 0.500 | 0.000 | 0.993 |
| | Weather | 0.000 | 0.900 | 0.250 | 0.000 | 0.925 |
| User | Physical Attack | 0.500 | 0.500 | 0.000 | 0.800 | 0.950 |

**FIGURE 5.4**

Example of resilience calculations by individual threat.

## 5.1.1 Individual Threats

The equation in Figure 5.2 is used to calculate a resilience value for each threat or threat scenario (which can include multiple threat types). If there are multiple threats, there is a resilience value for each, calculated separately for each threat. This results in a table of resilience values according to each individual threat similar to that shown in Figure 5.4.

## 5.1.2 Multiple Coordinated Threats — Cumulative Impact

In the case of multiple independent *but coordinated* threats, executed simultaneously, a further calculation can be made to find a cumulative resilience based on Equation (5.7):

$$R_s = R_1 - \sum_{n=2}^{N}(1 - R_n) \tag{5.7}$$

Where:
  $N$ = Number of threats
  $R_1$ = Resilience to first threat
  $R_n$ = Resilience to the $n$th threat
  $R_s$ = Cumulative system resilience for $N$ threats

It is important to establish that multiple threats do not impact the same parts of the system or else capability loss will be double-booked, as it can only be lost once. An example of the result of this calculation is shown in Figure 5.5, in which it is clear that the resilience monotonically decreases as additional threats are added (so long as the resilience values are less than 1).

$R_n$

| Segment | Threats | Avoidance | Robustness | Recovery | Reconstitution | Resilience | Cumulative Resilience |
|---------|---------|-----------|------------|----------|----------------|------------|------------------------|
| Space | RF Jamming | 0.500 | 0.100 | 0.850 | 0.000 | 0.933 | 0.9325 |
| Ground | Physical Attack on Ground Station | 0.500 | 0.750 | 0.600 | 0.200 | 0.960 | 0.8925 |
| Space | Nuclear Near Burst | 0.000 | 0.900 | 0.900 | 0.100 | 0.991 | 0.8835 |

$R_s$

**FIGURE 5.5**

Example of cumulative resilience calculation.

| System Segment | Threat Probability $P_n$ | Threats | Avoidance | Robustness | Recovery | Reconstitution | Resilience $R_n$ | $(1-R_n)P_n$ Weighted Loss |
|---|---|---|---|---|---|---|---|---|
| Space | 0.75 | Jamming | 0.00 | 0.75 | 0.15 | 0.00 | 0.79 | 0.159 |
| Ground | 0.20 | Physical Attack on Ground Station | 0.85 | 0.80 | 0.25 | 0.25 | 0.98 | 0.003 |
| Space | 0.05 | Nuclear Near Burst | 0.20 | 0.45 | 0.90 | 0.35 | 0.97 | 0.001 |

Total expected loss 0.164
Probability-weighted Resilience ($R_{PW}$) 0.836

**FIGURE 5.6**
Example of probability-weighted resilience calculation.

## 5.1.3 Probability-Weighted Resilience

Deriving a single resilience inclusive of the effects of all threats can appear attractive but usually yields less insight than considering a set of values. Nevertheless, there are mathematical techniques that can provide such a metric, though the result does not necessarily represent a real-world scenario (e.g., there is no assumption that the threats are coordinated). One approach that was previously discussed in Chapter 2 is shown below and requires one additional piece of information: an assessment of the *relative* probability of a threat occurring within a time period of interest. These relative probabilities, $P_n$, must sum to 1 in order to preserve a resilience range of 0 to 1. This calculation, sample results of which are shown in Figure 5.6, weights the resilience value for each threat by its accompanying probability of occurrence to produce a "weighted" resilience value that encompasses the range of threats to which the system may be subjected. This method scales the resilience values for each threat by the probability of the threat occurring, thus giving greater value to more probable threats. Equation (5.8) shows this calculation:

$$R_{PW} = 1 - \sum_{n=1}^{N}(1-R_n)P_n \qquad (5.8)$$

Where:
$R_{PW}$ = Probability-weighted resilience ($N$ threat scenarios)
$R_n$ = Resilience to $n$th threat
$P_n$ = Relative probability of occurrence for $n$th threat scenario

## 5.1.4 Superposition of Threats

There is another approach to multi-threat analysis that can also prove useful. If the individual system responses to each threat are known, then the response of a coordinated threat can be found by superposing the responses, assuming independent events. This approach is shown in Figure 5.7. Here the system responses to two threat events are known. These time domain system responses can be combined through superposition to

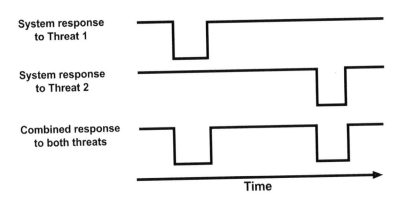

**FIGURE 5.7**
Superposition of system responses to create a combined response.

produce a combined response, which can then be modeled and analyzed as a single event. Modeling of concatenated threats is discussed in greater detail in Chapter 6.

## 5.2 Determining the Resilience Coefficient Values

The difficulty in deriving credible and accurate values for the attributes in the resilience equation varies according to the attribute. The most straight-forward estimate is likely that for robustness, which is similar to other engineering analyses performed to predict system behavior. Similarly, the recovery metric can be well estimated if the system behavior is well characterized based on design features and the concept of operations (CONOPS). Since the options for reconstitution are somewhat more limited than for the other methods of achieving resilience, that metric might be easier to determine, if only due to fewer available options.

The probability of avoidance is the most difficult and potentially subjective value to determine. It largely depends on the accuracy and depth of the knowledge about the nature of the threat or adverse condition as well as the available mitigations. It can include a wide range of input data, including political assessments, leading to an estimate of deterrence, for example. There are many aspects to avoidance, and it can be difficult to fully capture and validate all that are relevant and credible. Each resilience attribute is more fully explored later in this chapter.

The resilience estimate is also a snapshot in time, based on best available data informing the metric values during the design phase. As new information is received and as conditions change the assessments must be adjusted accordingly, resulting in updated resilience values. It is clear that resilience is

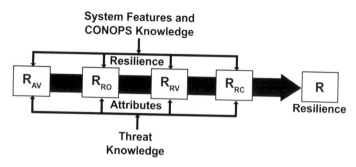

**FIGURE 5.8**
Resilience equation coefficient values are informed by key knowledge.

not necessarily static. Figures 5.8 and 5.9 illustrate how the inputs propagate into the resilience coefficients and how, over time, the estimated resilience value can vary as the understanding of the threat, the system, and the environment is refined or changes.

## 5.3 Modeling Resilience Attributes

### 5.3.1 Avoidance

Methods of increasing avoidance are among the first items to be considered when trying to impart resilience to a system. If the threat can be avoided, no other steps must be taken. There is no need to worry about how robust the system is, or whether it can recover or be reconstituted. Avoidance is

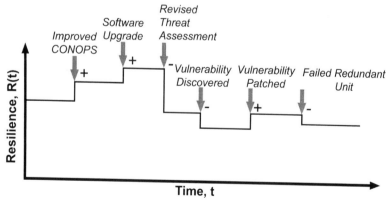

**FIGURE 5.9**
Resilience as a time-varying function.

sufficient. Avoidance is defined such that it is the probability that *no* capability is lost as the result of a threat. If any capability is lost, the loss is accounted for by the robustness parameter.

There are many methods of implementing avoidance in a space system, and these depend on the nature of the threats being mitigated. Referring back to the threat mitigation timeline in Figure 4.1, activation of avoidance features requires some identification, detection, and monitoring of the threat through indications and warnings that alert the system that an incoming threat is imminent. The earlier the warning, usually the higher the probability of avoidance. There are two possible ways to accomplish this. In the first case, the warning comes as an input from an external system. In the second case, the system itself is capable of detecting an inbound threat and activating the necessary avoidance mechanisms. The result might be the same, but in one case the system designer need not worry about incorporating additional threat detection apparatus, however, the resilience of the indications and warnings (I & W) system must also be considered, while in the other case the I & W functionality must be included in the system design.

Avoidance is a broad term, encompassing any number of methods for achieving it. For natural phenomena such as solar flares, historical data and near-real-time observation might be the best way to evaluate the threat and determine which avoidance methods will be effective. When dealing with hostile adversaries, the avoidance assessment becomes more complex. The threat assessment can include such disciplines as political science, game theory, and even psychology. The threat assessment is now based on the motivations, plans, and proven or suspected capabilities of one or more adversaries. First and foremost, how capable is the adversary of exercising the threat? Is the adversary a near peer or a small non-state group? If a global power or nation state, how does politics enter into the equation?

Modeling avoidance is not quite as straightforward as it might seem. This is because avoidance occurs in the pre-event region of the timeline and it is vital to define the exact beginning of this time period. In Chapter 4 a notional threat mitigation timeline was presented (Figure 4.1) with the time period of interest commencing with the initiation of the threat action, which sometime later might result in an event, representing the moment of impact on the system. Some have proposed an even broader definition, which includes gauging the *intention* of the adversary, leading to a likelihood of that adversary initiating the action. While this might seem like an exercise in semantics, many consider this to be an essential aspect of threat assessment.

The reasoning is that the greater the apparent resilience of a system, the less likely an adversary is to attempt to threaten it. This is the principle of deterrence. If the system is highly resilient, or its design presents the adversary with a poor benefit to cost ratio, meaning the costs of attacking are greater than the apparent benefits, then there is less chance of an attack being initiated by rational adversaries.

This definition may seem in conflict with a prior assertion, namely that the evaluation of resilience does not depend on nor incorporate the likelihood of a threat actually occurring. While this could be considered a gray area, including deterrence does not make any presumption regarding the likelihood of occurrence in any given time period, nor its frequency of occurrence. It's a fine distinction, but one worth mentioning. In any event, leaving deterrence in the model or taking it out is ultimately a design choice, at least partially based upon the confidence of the value of such an assessment.

Additionally, there is another possible mitigation option prior to initiation. If the threat can be monitored, and there is ample warning of imminent activation, then it may be possible to preempt the initiation of the threat. Preemption would then be yet another component of avoidance. An example would be if an adversary's ASAT launch complex was monitored and an intelligence system provided warning of an upcoming launch in time for the launch to be preempted.

The benefit of avoidance is illustrated in Figure 5.10, which shows the increase in system resilience over a threat range as the probability of avoidance for each individual element in an eight-element system increases from 0 to 0.95. For simplicity, this notional system has no elemental robustness, and no recovery nor reconstitution features. In this example, one or more of the eight elements, each contributing equal capability, are targeted by the threat. The number of elements targeted is increased parametrically over a threat range T. If an element is targeted and the threat is not avoided,

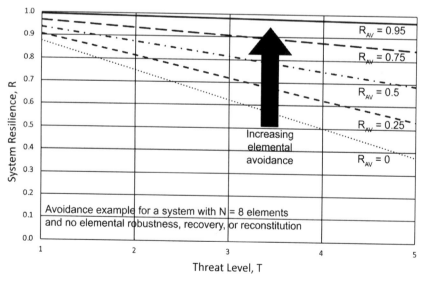

**FIGURE 5.10**
Impact of avoidance on system resilience.

it is permanently disabled, meaning the robustness of each is zero (this also assumes threats with perfect effectiveness, but more on that later). For a baseline case with no avoidance ($R_{AV} = 0$) for each additional threat the resilience drops by one-eighth, as an eighth of the total system capability is lost (R will eventually become zero for T = 8). As each element's avoidance to the threat is increased, the system level resilience slowly rises with it: For a 50 percent probability of avoidance for each element, R climbs to between 0.94 (T = 1) and 0.69 (T = 5), a significant improvement. For a threat level of T = 4, the system resilience improves from 0.5 (four of eight elements lost) to as high as 0.98 for an avoidance of 0.95 for each element.

A sample avoidance model is shown in Figure 5.11. This model includes parameters for both deterrence and preemption probabilities prior to the threat being exercised. If these are not present, their value is zero. The next parameter is the probability of effectiveness of the threat itself. As noted in Chapter 3, the probability of the threat's effectiveness, independent of any countermeasure, is often referred to as $P_k$, or probability of kill. No real-world threat is 100 percent effective and any assessment of $P_k$ should be included in the avoidance value if known.

The model also includes countermeasures against the threat. The intended effect of a countermeasure is to mitigate a threat by reducing the threat effectiveness by counteracting its effects. This resulting value can be expressed as an "adjusted" or "residual" $P_k$. An indications and warnings feature (or system) is required to detect the threat and thus activate the countermeasure mechanism early enough for it to be effective. This involves two additional parameters: the probability that the incoming threat is detected (and tracked, if necessary) and the probability that the countermeasure itself is effective. Just as the threat is not 100 percent effective, neither is the countermeasure. These probabilities combine to form the avoidance parameter, as shown in

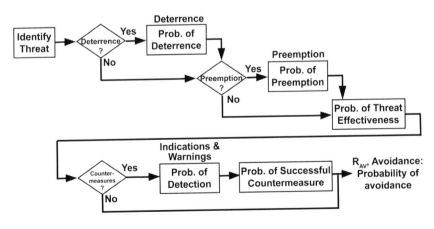

**FIGURE 5.11**
Sample avoidance model.

| Variable | Avoidance Component | Value |
|----------|---------------------|-------|
| $P_{DT}$ | Probability of deterrence = | 0.50 |
| $P_P$ | Probability of preemption = | 0.10 |
| $P_E$ | Probability of effectiveness = | 0.95 |
| $P_D$ | Probability of detection = | 0.75 |
| $P_C$ | Probability of counter = | 0.90 |
| $R_{AV}$ | **Probability of Avoidance =** | **0.86** |

**FIGURE 5.12**
Sample avoidance model calculation.

Equation (5.9), along with a sample tabular format with notional parameter values, shown in Figure 5.12.

$$R_{AV} = P_{DT} + (1 - P_{DT})P_P + (1 - P_{DT})(1 - P_P)(1 - P_E) + (1 - P_{DT})(1 - P_P)P_E P_D P_C \quad (5.9)$$

Analyses of both the threat and the system are required to produce the parameter values. As such information becomes available, this method provides a framework which can be used to make the calculations that enable comparison among multiple alternatives. Any parameter can be omitted from the model simply by setting its value to zero.

The model also reinforces certain fundamental principles. As with the capstone resilience equation, there are multiple ways to achieve avoidance that can be equally effective. If a threat can be deterred with 100 percent certainty, the result is as effective as 100 percent certain preemption or a 100 percent effective countermeasure. The discriminator in choosing an avoidance approach is not necessarily the methods themselves but other considerations such as implementation cost. Once the need for improved resilience is recognized, each of these alternatives can be considered and assessed as to the likelihood of being successful as well as determining the associated costs.

## 5.3.2 Robustness

There are two aspects to robustness that must be considered. Robustness is first considered at the element level: satellites, ground stations, and other functional blocks. Then there is the matter of the robustness at the system level for systems that feature many elements.

Most space systems are very robust at the element level because the elements are built to very high-quality standards, including rigorous testing, particularly the space segment, due to harsh space environments and the lack of available repair options on orbit. Passive and active redundancy are used to provide this robustness as well as precision workmanship and verification. Protection techniques provide resistance to environmental conditions

**FIGURE 5.13**
Robustness model.

such as radiation (both natural and artificial), temperature variation, and solar activity (flares, wind, etc.). The simple model is shown in Figure 5.13.

For most space system threats, including electronic and kinetic, the loss of capability coincident with an event is nearly instantaneous and thus the simplified expression for robustness can be expressed as shown in Equation (5.10):

$$R_{RO} = \frac{C_e}{C_0} \qquad (5.10)$$

where $C_e$ is the level of capability that remains following the threat-induced event and $C_0$ is the pre-event capability level. If the capability is lost over some finite period of time during a sustained incident, this equation can be modified to take into account the profile for the loss of capability when calculating $R_{RO}$.

New threats require new techniques and technologies to improve robustness for the cases in which avoidance is unlikely. These protection measures can include features such as anti-jam techniques, incorporation of more robust sensors (both RF and optical), and high-dynamic range RF electronics with increased damage thresholds.

On the ground, physical fortifications can be increased, also with higher costs. Bunkers, barricades, armor plating, and more security forces are all examples of increased robustness to conventional attacks. Burying optical fiber deeper to dissuade sabotage can provide additional network robustness as can the installation of firewalls and other cyber protection.

Historically, the costs of adding robustness have risen significantly, particularly when incorporating large amounts of redundancy. This is motivation to find more affordable alternatives through avoidance and recovery.

Robustness takes on a somewhat different appearance when considered at the system level. Previous examples considered systems in which capability was delivered through several satellites to multiple ground stations. If a satellite is lost, its capability is also lost, and the system capability also suffers, often proportionally if each satellite delivers equal capability. In this case, the system-level robustness is a fraction of the original capability value. So even though the robustness of individual satellites might be very low, the system-level robustness can still be relatively high, depending upon the severity of the threat. System-level implications of robustness are discussed in further detail in Chapter 6.

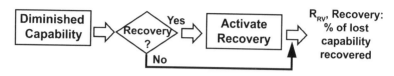

**FIGURE 5.14**
Recovery model.

## 5.3.3 Recovery

A simple recovery model is shown in Figure 5.14 wherein lost capability is restored through the activation of recovery features inherent in the system design. Recovery features can be active or passive, autonomous or due to human intervention ("person in the loop"), partial or full, and discrete or continuous, as shown in Figure 5.15.

Based on the DoD definition of resilience, *shorter periods of reduced capability* is a value metric for recovery. A system with shorter periods of reduced capability, as well as recovery of a greater amount of the lost capability should result in a higher evaluated value of resilience compared to systems or architectures that provide less of each.

The concept of recovery is straightforward. Ideally, a recovery feature is activated immediately after a threat event, resulting in full recovery of capability. For this special case of near instantaneous recovery, the value of $R_{RV}$ approaches 1. For all other cases, the calculation becomes more nuanced based on how the recovery function is valued and the choice of the recovery period. Two key considerations must be addressed for the purposes of the calculation: establishing the period of interest for evaluation of the metric, including normalizing the calculation time duration, and whether to establish a minimum acceptable capability threshold. Each one of these choices affects the resulting metric value.

Clearly the calculation of the recovery metric should be based both on the recovery time as well as the amount of the original capability recovered.

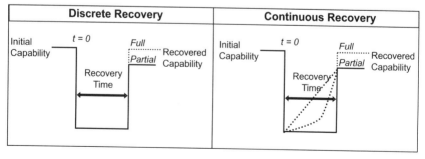

**FIGURE 5.15**
Discrete and continuous recovery profiles.

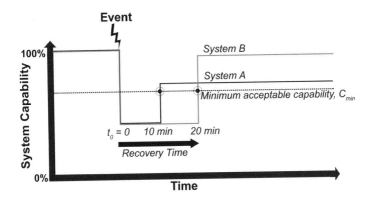

**FIGURE 5.16**
Simple recovery calculation based only on time.

This can be done through knowledge or estimation of the capability versus time function, as shown in Figure 5.16, and results in an equation of the form described in Figure 5.17. This equation derives the recovery coefficient $R_{RV}$ through an integration of the capability versus time function that has been normalized such that the initial steady state value is 1. A second function, $v(t)$, is also optionally included, and is a time-based value function which is discussed later in this chapter. This function provides a method of windowing the capability function and overlaying a weighting factor based on value to the user versus time.

One simple recovery metric calculation methodology is to measure or estimate the time at which the capability remains below some minimum value following loss of capability, assigning higher recovery values to shorter outage durations. For example, if System A regains its capability above some defined minimum capability value $C_{min}$ in 10 minutes, and System B does the same in 20 minutes, then values can be assigned such that $R_{RV}$ for System A is twice that of System B, and System A is advantaged in the overall resilience calculation (Figure 5.16). But if System B regains a *greater percentage* of the original capability above $C_{min}$ at 20 minutes, then no additional credit is provided, and this could disproportionately disadvantage it relative to user value. For this and other reasons, a careful choice of

$$\underbrace{R_{RV}}_{\substack{\text{Recovery}\\\text{metric}}} = f(C,v)(t) = \int_{t=0}^{T} \underbrace{C(t)}_{\substack{\text{Capability}\\\text{recovery}\\\text{vs. time}}} \underbrace{v(t)}_{\substack{\text{Time-based}\\\text{value function}}} dt$$

**FIGURE 5.17**
Notional form of recovery metric defining equation.

evaluation methodology should be made to ensure that the resulting values of $R_{RV}$ provide the desired comparative weighting to properly represent recovery value for each system design.

Using the equation in Figure 5.17, the start of the calculation period is the time at which system capability is first lost, concurrent with the event, shown as t = 0. The more debatable choice is the time at which to end the calculation period (time T). A simple case illustrates the effects of the choice of the calculation endpoint upon the resulting recovery value. Two different systems each suffer a complete loss of capability. System 1 recovers full capability in $T_0$, and System 2 in $T1 = (2)(T_0)$, after which both return to 100 percent steady state performance. This scenario is illustrated in Figure 5.18. An intuitive choice for the calculation endpoint is the time at which both

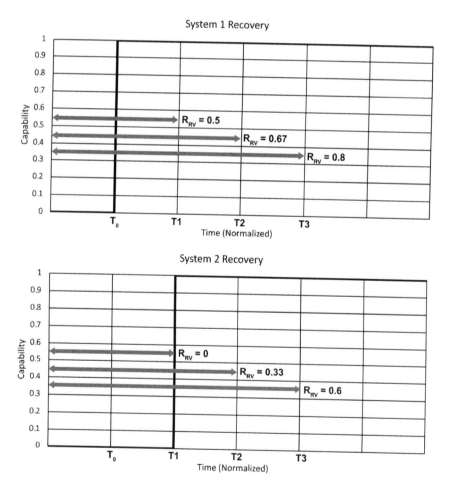

**FIGURE 5.18**
Effect of choice of recovery calculation period.

systems have regained steady state, in this case shown as T1. The calculated recovery value for System 1 is 0.5:

$$R_{RV} = \int_{t=0}^{T1} C(t)dt = 0.5$$

and for System 2 it is zero. This choice reflects an assertion that there is no recovery value after time T1 and no credit is received after that point.

If instead the calculation period ends at T2, a different valuation is made. Then System 1's recovery value is 0.67, and System 2's is 0.33. This reflects a somewhat intuitive result that System 1 is twice as resilient as System 2 since its outage time is half as long. This choice also reflects the assertion that beyond time T2 there is no further value to the user that is to be captured in the calculation. Finally, if the calculation is extended to time T3, the values increase again: System 1's recovery value is 0.8 and System 2's is 0.6. Note that these values are steadily converging.

The convergence for this case is because, as the end time T increases, the outage times relative to the calculation time become small, and the comparison becomes a small difference of large numbers. Put another way, if the recovery value calculation extends over a long time relative to the outage times, then small differences are reflected in the recovery metrics. A five-minute outage versus a ten-minute outage relative to a time period of interest measured in days is practically indistinguishable and the resulting calculation reflects that.

Now consider a second case, in which the recovery times of the two systems are the same as in the previous example, but System 1 only recovers 50 percent of the lost capability while System 2 fully recovers. If the same thought experiment is performed as before, as T becomes very large the second system's recovery value becomes twice that of the first because in the extreme the second system's value will approach 1 while the first will approach 0.5. Nevertheless, the first system returned partial value sooner than the second. So, does this valuation accurately reflect the relative merits of the two systems? This illustrates the importance of defining the period of interest from $t = 0$ to T such that the utility value of the recovery profile renders an accurate comparison between two or more system designs.

A convenient method for clearly defining the period of interest is to include a value or windowing function shown in Figure 5.19 as $v(t)$. Using a value function that weights the integrated value provides a windowing function for the purposes of the calculation. In the simplest case, it can be used to establish the point in time at which the integrated function is zero (Figure 5.19). If the value function is unity ($v(t) = 1$) over the period of interest, the resulting value of $R_{RV}$ is simply equal to the area under the capability curve over that time frame, forming a capability-time product, as shown earlier. But the value function can also be used to weight the recovery profile values to more

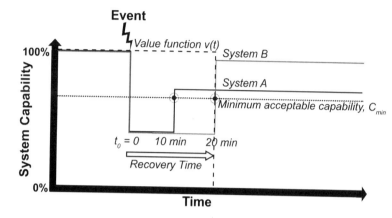

**FIGURE 5.19**
Adding a value (windowing) function.

strongly modulate the value of $R_{RV}$. Note that in Figure 5.19 the period of interest only extends to 20 minutes, after which any utility to the user is zero. System B recovers a greater percentage of capability but does so too late to be given credit for it. Had the value extended beyond 20 minutes, System B's recovery metric would increase relative to System A's.

There are two benefits to incorporating a value function. The first is simple: When used as a windowing function, it restricts the calculation period such that only the period of interest is considered. As previously discussed, if the calculation of $R_{RV}$ is extended for long durations after the recovery has reached steady state, the value increases, potentially distorting comparisons between different design architectures.

The second benefit of a value function is to enable more complex value weighting throughout the recovery period. In the previous example, the value of the windowing function was unity throughout the recovery period. A more continuously varying function can also be used, as shown in Figure 5.20. The use of a value function to adjust the raw recovery value is predicated on the idea that as time passes the recovered utility value of the system capability is less, and could be highly non-linear, or even discrete (zero value after some time $t$). The rationale for such an approach is the military concept of an "operationally relevant timeline."

The basic principle is that in a conflict, there is a period of time during which there is utilitarian value to the military for a system capability or availability of a resource. After some time, there is little operational value because the opportunity to use the capability to sway the outcome of the battle has already been lost. If this valuation of capability versus time can be assigned to a value function, this function can be overlaid, as shown in Figure 5.20, to adjust the value of the recovered capability accordingly. The fact that one system design or architecture delivers a greater amount of recovered capability

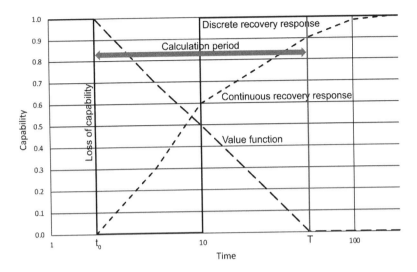

**FIGURE 5.20**
Recovery value as a function of time.

**Source: "A Method for Calculation of the Resilience of a Space System," MILCOM 2013, R. Burch, 2013.**

but does so at a later time can then be factored into the overall resilience evaluation. Regardless of the shape of the value function, every recovery evaluation calculation should include one to provide a basic windowing function to establish the calculation endpoint time, T. And values for $t_0$ and T are chosen to normalize $R_{RV}$ to take on a range of values between 0 and 1.

For these simple examples, no minimum capability threshold was established, though one can be added very easily. In this case, loss of capability is only counted if the value falls below this threshold for some period of time, and entails putting another constraint on the calculation, but this time for the $y$-axis value. This can be yet another functional overlay that enables accrual of recovered capability value only when the recovered value reaches this minimum threshold.

For modern space systems a significant portion of the recovery time might be due to human activities that limit how quickly service can be reestablished. This includes time to detect the issue, diagnose and attribute the cause and location within the system, and then obtain approval to enact the recovery procedures. Operators are extremely careful regarding satellite commanding since mishandling could result in loss of the asset. Satellite telemetry is often the only information available to operators to diagnose issues and the process can take some time. The use of pre-coordinated autonomous recovery mechanisms is thus desirable whenever possible. This includes the use of autonomous passive redundancy switching where allowable. Active redundancy, which does not require switching, is an "operate through" condition

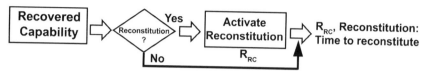

**FIGURE 5.21**
Reconstitution model.

and therefore a method associated with robustness as opposed to recovery, as this method prevents the loss of capability in the first place.

Some autonomous systems take a period of time to achieve full performance, such as iterative adaptive nulling antenna algorithms for anti-jam applications. In this case, there is some time period where communications are degraded or inhibited prior to the antenna system adapting and the algorithm converging to provide interference suppression.

### 5.3.4 Reconstitution

Reconstitution differs from recovery in that reconstitution results in the replenishment of full capability for *all* services or missions whereas recovery applies only to a subset. While partial recovery is possible, partial reconstitution is, by definition, not: Following successful reconstitution, *the original capability has been restored*. This is the discriminating characteristic of this mitigation approach, and its value is greatest when the other attributes: avoidance, robustness, and recovery, have failed to provide sufficient resilience.

Modeling the value of the reconstitution metric may be more subjective than for the other resilience attributes. As the final capability value is, by definition, 1, the only variable truly in play is the time to reconstitute, treated similarly as with recovery. As reconstitution is assumed to take longer than any recovery features, the question is how to value the time element. As with recovery this is dependent upon the relative value to the user. But that value might also be influenced by the relative value of reconstitution versus the value of other methods of mitigation (and their costs).

The launch of a new satellite to replace one that has been disabled is but one example. Figure 5.21 shows the simple model for reconstitution, and Figure 5.22 shows a notional form of a calculation for a metric, similar to

$$R_{RC} = f(C,v)(t) = \int_{t=0}^{T} C(t)v(t)\,dt$$

Reconstitution metric        Capability recovery vs. time    Time-based value function

**FIGURE 5.22**
Notional reconstitution equation.

that of the recovery metric. Though the recovery response can be modeled as either full or partial, by definition reconstitution recovers all of the original system capability.

The primary value parameter for this attribute is time to reconstitute and again shorter is better. Based on the definition, either the capability is fully reconstituted or it is not, with the only other factor the time to reconstitute, which can be used to normalize $R_{RC}$ to ensure a value between 0 and 1.

Though the earlier timeline shows reconstitution occurring after recovery, theoretically it could be possible for reconstitution to occur first, obviating the need to recover, as reconstitution by definition results in a higher level of capability than partial recovery. In practice, for space systems it is more difficult to imagine being able to implement reconstitution more rapidly than a lesser recovery technique. A potential example might be the re-commissioning of a dormant satellite following the loss of capability on another satellite. This would likely be the exception and not the rule.

Implementation features to enable reconstitution in a timely manner can include redundancy or sparing at a very high level, highly responsive deployment systems for space, ground, and user segments, and perhaps cooperative operations among multiple systems at the enterprise level. In some cases, there might be multiple ways to provide the mission capability by using different systems to restore operations.

## 5.4 System Availability and Resilience

System availability is a measure of the percentage of time that an operational system is successfully delivering its capabilities. This metric is a high-level measure of the reliability of a system, and for space systems is sometimes referred to as *operational availability*. Availability $A_0$ is defined by Equation (5.11) [11]:

$$A_0 = \frac{Uptime}{Uptime + Downtime} \tag{5.11}$$

where the *uptime* is the time over some specified period that the system is operational, and the *downtime* is the time during which the system is performing below its required performance level or suffering an outage. This can be due to a variety of causes, including internal system failures, maintenance events, and calibration cycles. Availability can be further expressed in terms of mean durations as shown in Equation (5.12):

$$A_0 = \frac{MTBF}{MTBF + MTTR} \tag{5.12}$$

where MTBF is the Mean Time Between Failures, and MTTR is the Mean Time to Repair. This form of the equation is directly derived from system reliability through Equations (5.13) and (5.14):

$$MTBF = \frac{1}{\lambda} \tag{5.13}$$

$$R_{el} = e^{-\lambda} \tag{5.14}$$

where $\lambda$ is the aggregate system failure rate and $R_{el}$ is the system reliability. From these equations it is clear that both reliability and availability are based on *rates,* and thus average measures per unit time. This is important when considering the potential incorporation of resilience into system availability.

Resilience as defined previously in Chapter 2 and earlier in this chapter is a measure of the response of the system given one or more defined threats. While the resilience calculations provided in this chapter can estimate predicted outage times representing MTTR, the mean frequency of occurrence yielding MTBF is not a calculated value. As such it becomes difficult to incorporate resilience as defined into system availability *unless the frequency of threat occurrences can be expressed in terms of a mean rate.*

The aggregate failure rate of a system based on the failure rates of its constituent components can be found by Equations (5.15) and (5.16):

$$\lambda_s = \lambda_1 + \lambda_2 + \lambda_3 + \cdots \tag{5.15}$$

$$MTBF = \frac{1}{\sum\limits_{i=1}^{n} \lambda_i} \tag{5.16}$$

where $\lambda_s$ is the system failure rate and $\lambda_n$ represents the failure rates of the constituent components assuming that the individual components fail independently [12]. The aggregate MTBF is shown in Equation 5.16 [12]. In this construct, the contribution of resilience outages to the system availability could be represented by yet another term, $\lambda_r$, resulting in an adjustment to the original MTBF that is based solely on reliability contributors. Figure 5.23 illustrates how combining the outages due to reliability and resilience effects could describe an adjusted overall system availability.

This approach has several drawbacks. Historical data can be used to estimate the mean frequency of some adverse conditions such as debris collisions or solar flares but estimating the mean time between hostile actions can be difficult. Additionally, assumptions regarding the independence of outages and the potential for concurrent outages due to system failures and external forces can make this calculation a bounding case that underestimates the MTBF value. Though imperfect, examining the relative projected

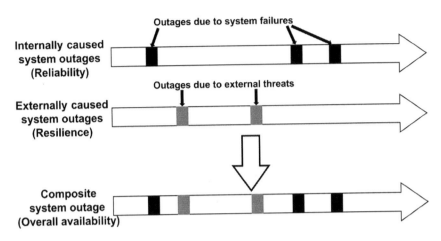

**FIGURE 5.23**
Incorporating resilience effects into system level availability metric.

outage times and frequencies for both reliability-driven events and resilience-driven events will provide the designer with a more complete picture of the overall system availability even if the calculation of an all-inclusive system level availability value proves to be impractical.

# 6

## Designing for Resilience

The design of a satellite system begins with the establishment of requirements for cost, performance, and resilience. As the system design is developed, resilience requirements are decomposed and allocated among the various system elements based on the defined threats. During the design process threat mitigation features are considered and traded to ensure that the resulting system meets these requirements. These design trades are crucial in optimizing the system design for cost, performance, and resilience. This chapter presents many key design approaches and trades that may be employed to create such an optimized system.

## 6.1 Establishing Requirements

The design process starts with the development of the system performance, cost, and resilience requirements. These requirements include the capabilities of interest and the level at which each is delivered. Examples for a SATCOM system include such parameters as communications capacity, data rates, error rates, and other performance measures. These requirements are first established at the system level and then allocated to major segments, subsystems, and eventually down to individual units that comprise the system. A cost target or budget is also established based on affordability as well as the required system capability and functionality. The resilience requirement is levied based on the credible and qualified threats to the system.

The next step is the development of candidate designs and the execution of trade studies to understand the interdependencies and sensitivities of any number of parameters to create a design that meets all requirements to the greatest extent possible. The analytical tools presented in Chapter 5 are used to perform these trades to optimize the system design. The process is iterative, as optimization may not be straightforward. Understanding the coupling and dependencies between the requirements can reduce the total design time required.

Top-level requirements may have to be revisited if cost targets are exceeded or required performance cannot be achieved. Lower-level requirements may be reallocated as part of the optimization process while keeping the higher-level requirement set largely intact. However, throughout the entire process the requirements must be kept at the forefront to ensure optimization is achieved to the greatest extent possible.

## 6.2 Incorporating Resilience Engineering into the System Design Process

As with other system attributes, resilience is best considered during the earliest stages of the design process as opposed to being relegated to an afterthought. There is the temptation to focus on the core system functionality and performance and leave requirements such as resilience for later. History has proven this to be an inefficient approach. Rather, resilience should be incorporated into the traditional system design process, as with other requirements such as reliability. As the fidelity of the design matures, so do the resilience design features. The designer must continue to balance cost, performance, and resilience throughout the entire process.

Figure 6.1 shows design for resilience as part of the standard systems engineering design process. The first step is defining the system functional and performance requirements, performing a functional decomposition and system model, and developing one or more competing architectural designs. The system requirements include the environmental conditions over which the system must function and also identify the threats that form the contested operational environment. A threat assessment is performed to determine which are truly of concern and are credible in their ability to impact system performance. In the final step, the resilience of each of the candidate designs is assessed and compared.

A key step in the process is the evaluation of the threats. The goal is to ensure that only the threats that are design drivers are considered. How this

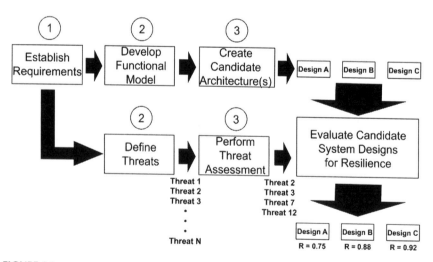

**FIGURE 6.1**

Evaluation of resilience as part of the systems engineering design process.

**Source:** "A Method for Calculation of the Resilience of a Space System," MILCOM 2013, R. Burch, 2013.

screening is performed may vary. One accepted method is to perform a risk assessment by estimating the likelihood (L) (or probability) of the threat being successfully exercised, combined with the consequence (C) of it occurring, as discussed in Chapter 2. All identified threats are evaluated to determine which of them exceed a predefined L × C threshold. Only threats meeting these criteria are considered during the design phase. This approach eliminates the design from being unduly influenced by threats that are either highly unlikely or have negligible consequence to the system operation if realized.

While it is highly desirable to distill the resilience of an architecture across all threats to a single number, in practice this is difficult if not impossible to do. It may also be undesirable, as an aggregated number can obscure the nature of the driving threats when competing architectures are scored and evaluated. Situations in which some simple methods of developing cumulative resilience values can be performed are illustrated in a later section of the chapter. It is much more likely that the result of a complex resilience evaluation is a table of values detailing the resilience of the system to each threat for the multiple capabilities of each architecture considered. Prioritization of the threats and capabilities may then guide designers, as well as the relative likelihood of the specific threats being executed and the mitigation costs for each.

### 6.2.1 Architectural Trades

As discussed in Chapter 1, the architecture of a system is directly related to the means by which capability is delivered to its users. It is therefore appropriate to begin with the system architecture when considering how to bolster system-level resilience. Often the process begins when one or more competing architectures have been developed and all credible threats have been identified and characterized. The system's concept of operations (CONOPS) have also been established, meaning that the way in which the system is operated is understood. Prior to the start of the design process, a specification or goal for the maximum allowable cost and the minimum system performance and resilience have been established.

The next step in the design process is to assign the threats to the targeted elements of the system. If the system appears to be at risk of not meeting its resilience goal, the designer may choose to allocate resilience to the key system elements to ensure it does. This is done by identifying the areas that require threat mitigation features to achieve a higher level of local resilience and performing an analysis to determine the local resilience values required. Increased redundancy of vulnerable elements is an example of this approach.

Adding selected protection and/or tailoring of the architecture to eliminate the vulnerable elements may accomplish this. An example of tailoring an architecture to remove vulnerabilities is strategic MILSATCOM (military satellite communications), in which fixed ground sites are considered high-value targets in a nuclear conflict, and unlikely to survive a nuclear attack regardless of location or protection level. As a result, the modern strategic MILSATCOM system

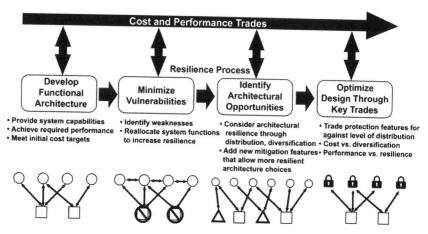

**FIGURE 6.2**
Representative architectural design process.

does not include such vulnerable targets, instead locating considerable digital processing in the nuclear hardened space segment and employing a few highly mobile and distributed command elements in the air and ground domains.

A representative architectural design process is presented in Figure 6.2. The first step, as always, is to define the baseline functional system architecture that is capable of meeting the top-level requirements to deliver some defined capability. Initial cost and performance goals are established at this time. The next step, with knowledge of the threat environment in hand, is to identify any obvious vulnerabilities that may be removed through changes in the architectural design. Next, the designer should look for opportunities to add resilience through distribution and diversification within the architecture, if required. Finally, further detailed trades may be completed to balance distribution, diversification, elemental protection, and other measures to further optimize for cost and performance.

## 6.2.2 Elemental Protection

Historically perhaps the most fundamental way to protect space assets has been providing them with some manner of built-in protection against specific threats, or elemental protection. These threats could include solar flares, space debris, or radiation effects from nuclear explosions. The spacecraft design incorporates features that enable it to survive and recover or operate through the event or environment. The cost is generally highly proportional to the level of robustness (and thus resilience) required. For very high levels of robustness, the cost can be very high. Combining elemental techniques with architectural mitigations can provide the system designer with a wider range of design solutions for resilience.

One of the more significant issues is the possibility of threat evolution and escalation which may render one or more threat mitigation features ineffective, with little recourse to make changes to the satellite's posture on orbit. A second issue is that often specific threat mitigation features are designed to mitigate only one of multiple threats. This means that a satellite might require multiple mitigation features, with the commensurate increase in cost. In contrast, system-level mitigations such as distribution can address both of these issues albeit at an added cost as well.

### 6.2.3 Distribution of System Capability

Another method of providing system-level robustness is through the partitioning of the total system capability among a number of system elements, often referred to as *distribution*. As most space systems are to some degree inherently distributed, the focus is on *distributing mission capability beyond that which would otherwise be required to meet the minimum system performance requirements for the purpose of improving system-level resilience*. A highly distributed system achieves its resilience through system-level robustness. Even though the individual system elements may not be robust against one or more threats, the system continues to deliver its capability because the level of distribution allows it to tolerate some percentage of element losses while still delivering acceptable levels of capability. This property is sometimes referred to as *graceful degradation*, meaning that there are no abrupt declines in capability as elements are incrementally lost.

Distributed systems possess attributes that can result in both increased resilience and reliability and can provide graceful degradation from both types of capability losses. One of the key goals of providing protection against a hostile adversary is reducing the number of high-value assets (HVAs) that can become attractive targets. A distributed system presents a greater number of targets of lesser value, tilting the benefit to the cost trade in the system operator's direction and discouraging an attack (reducing motivation). Furthermore, these architectures can allow for more agile and complex CONOPS, complicating an adversary's planning.

Distribution is a very attractive means of increasing system-level resilience, in part because *this single method can provide resilience against a wide range of threats, independent of their nature*. A highly distributed architecture provides resilience against physical attacks, electronic attacks, natural disasters, and the like by presenting multiple targets and limiting the amount of capability lost when one or more elements become disabled. Many other threat mitigation techniques only provide resilience against a specific threat.

There are several operational distributed space systems, but the Global Positioning System (GPS) is a particularly good example. The base GPS system consists of 24 satellites in multiple planes in Medium Earth Orbits (MEO). Today, the system fields more than 30 satellites, providing on-orbit redundancy in the case that one or more were to fail or be disabled. This is a highly resilient system to many threats. As the location accuracy, a key capability, is proportional to the number of satellites in view, there is graceful degradation to navigation accuracy as the number of available satellites is reduced.

### 6.2.3.1 Limits of Distribution

Among the first questions to ask in considering distribution as a mitigation approach is whether the nature of the capability and the manner in which it is delivered constrains the ability to segment and distribute it. Establishing practical cost and performance limits based on the system architecture and implementation options results in defining a performance floor and a cost ceiling. The designer should establish these bounding limits as soon as is practical to determine the extent to which the capability can practically be distributed.

Figure 6.3 illustrates the relationship between number of elements and the amount of capability per element for the simple case of a fixed total capability and N elements with identical amounts of this capability. This is an inverse relationship mathematically. The resilience calculated is for a loss of a single element in a system of N uniform elements and is defined by $R = (N - 1)/N$. Starting with a unity capability of 1, the capability C per element decreases with N and is expressed as $C = 1/N$.

The first thing to note is that for the case of the loss of a single element as shown in the figure, the resilience asymptotically approaches 1 as the element count increases. The resilience obtained by choosing a system with 10 elements achieves a resilience of 0.9. Beyond that point the resilience gain becomes asymptotically small. Eventually there are diminishing returns as

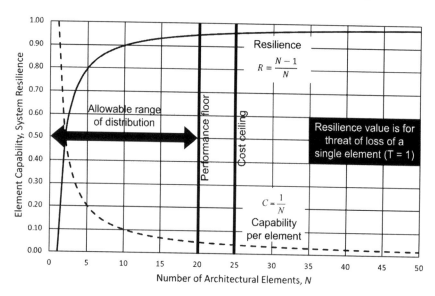

**FIGURE 6.3**
Capability per element and resilience versus element count.

the expense of adding elements begins to outweigh the added resilience that they provide.

This figure also illustrates how cost and performance constraints may limit maximum level of distribution by placing a maximum value on the number of elements. At some point the size of each individual element may shrink to the point that some minimum capability level cannot be sustained by that size of element, establishing the performance constraint. Likewise, as the aggregate system cost is proportional to the number of elements, at some point the cost becomes the practical constraint. Which of these becomes the limiting factor depends on the specific system. In this example, the performance floor is reached prior to the cost ceiling and thus becomes the dominant constraint on the maximum degree of distribution. It is also possible that the cost of the incremental resilience increase may constrain the element count even sooner. Or the designer may choose to simply meet the minimum resilience requirement. In this case if the minimum resilience is 0.9, the design only requires 10 elements.

Some capabilities can be more easily partitioned than others; in some cases, the capability that must be delivered severely limits the minimum permissible element size. For example, communications capacity, measured by total available bandwidth, can be imagined to be partitioned both uniformly and non-uniformly among a wide variety of satellites and payloads of varying sizes. On the other hand, a sensing capability may be much more difficult to partition as the sensor or instrument required to produce the images at a particular resolution may require a host spacecraft with a minimum size, weight, and/or DC power and fixed aperture size.

A simple example illustrates the potential effect of these constraints. Consider a SATCOM system that must close a certain high data rate link operating at a specific bit error rate (BER) into a given coverage area. The required link performance drives the size of the antenna and the transmit power on the spacecraft. A spacecraft's size, weight, and DC power thus limits the maximum data rate that can be sustained. But element size may also limit the maximum *capacity* that can be delivered as well, particularly with regard to available RF transmit power. At some point the spacecraft size cannot support the required antenna size or transmit power to close the link, and the capability is thus not delivered. Data rate (or capacity) thus becomes the constraining factor for minimum element size (in terms of capability delivered) rather than the capability that can be placed on the platform of a given size.

The cost constraint can be equally nuanced. Space system costs are often dominated by the number and size of the satellites as well as the accompanying launch vehicles. Distributing capability among a large number of smaller satellites may result in manufacturing cost savings due to higher volumes, but this savings may be offset by the greater number of launches required to put them in orbit. Thus, the net cost may be higher or lower than a less distributed system. As a result, it is crucial to make fair comparisons using life-cycle costs or total cost of ownership.

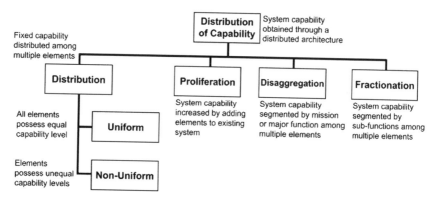

**FIGURE 6.4**
Methods of distributing system capabilities.

## 6.2.4 Methods of Distributing System Capability

There are many different ways to parse out a capability (Figure 6.4). The simplest way is to divide the total capability uniformly among some number N elements, such that for some threat level T, some number of elements may be lost while still retaining a minimum acceptable level of capability. A variation is to divide the capability in a non-uniform manner among those N elements. Or, given a number of uniform (or non-uniform) elements N, simply adding one or more additional elements of the same size to provide oversupply, which is referred to as proliferation. Disaggregation is the separation of functions within each element used to support multiple missions into discrete sub-elements to increase resilience. Fractionation is the segmentation of sub-functions providing one or more capabilities within an element. Each of these may affect the system-level resilience.

Consider the simple case of distributing a fixed capability among N elements, either uniformly or non-uniformly, with the goal of providing greater system-level resilience. Either of two similar methods can be used: distribution or proliferation. Distribution is partitioning a fixed amount of capability *optimally* among a number of elements, whereas proliferation is often expressed as the increase of capability by simply *adding* more of the same kind of element to increase system margin.

This is an important distinction because proliferation is often among the more expensive ways to obtain a distributed system since cost is usually proportional to total amount of capability and is increased by adding elements of fixed size and cost. When a capability is optimally distributed, the amount of excess capability required can be minimized, at least theoretically. In the case of proliferation, this is not always true. In addition, proliferation is sometimes more suited to the alteration of an existing system, whereas distribution may be best optimized when the system is in the design phase.

Choosing to distribute a capability does not relieve the designer from providing some amount of margin in the system simply due to the chosen architecture. For a given threat, there is a non-zero probability that there is some loss of capability, and the intended effect of distribution is to minimize that loss for some number of elements that are at risk. If that projected capability loss is greater than the minimum system requirement, additional margin must be provided to offset that loss.

In a distributed system, the manner in which the capability is distributed is an important consideration. Consider a system with N elements, and a capability value of C, which is also equal to the minimum required capability. As there is no inherent design margin, if even a single element is lost, the minimum capability requirement is not met. In this case, simple distribution of the capability (C) does not provide any benefit, as even if the capability is split into 2N or 3N elements, loss of *any* element results in an unacceptable capability level. Thus, the amount of capability must be increased prior to distribution to ensure the minimum value is met. As capability is proportional to cost, the question is how to most optimally add the minimum amount of additional capability to achieve the desired resilience.

Consider a SATCOM system that must supply some amount of global capacity, C. An identified threat is projected to result in the loss of a single satellite (T = 1). The minimum number of satellites to meet global coverage requirements is 4. If the capacity C is simply distributed equally across N satellites, then losing a single satellite results in some global capacity less than C. In this case, the *required* resilience is 1, since there can be no acceptable loss, and thus the required resilience cannot be achieved with a capacity of C. One solution is to modify the system design to increase the total capacity to a higher value, $C_N$, which is then equally distributed as before. The required resilience is now $R = C/C_N < 1$.

For example, for the requirement C = 10 Gbps, adding 2 Gbps for a $C_N$ = 12 Gbps yields a required resilience of R = 10/12 = 0.83, or 83 percent (Figure 6.5). This new total capacity ($C_N$) must now be distributed over N satellites such that

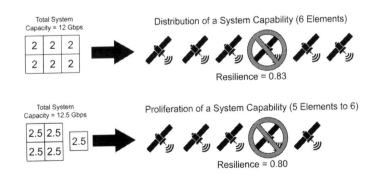

**FIGURE 6.5**
Comparison of distribution and proliferation.

losing one will result in less than 17 percent loss of capacity. Thus, $1/N < 0.17$. So, $N > 5.88$. This means that a minimum of six satellites is needed to meet the minimum required resilience (R). For six satellites (2 Gbps capacity each), loss of one satellite results in resilience of $R = 5/6 = 0.83$, corresponding to the required capacity of 10 Gbps. An alternative would be to grow each of the identical satellites while maintaining the original number of satellites such that the loss of one still meets the resilience requirement, however, there may be an upper limit on the size and capacity of the satellite, necessitating a distributed approach.

A proliferation approach provides a useful comparison. As in the previous case, the 10 Gbps is distributed equally to meet the requirement. Distributing among the minimum of four satellites yields 2.5 Gbps of capacity per satellite. This time, surplus capacity is added by simply *adding one more of the same satellite.* This results in a new system with $C_N = 10 + 2.5 = 12.5$ Gbps. This creates a greater oversupply than in the first case, and the required resilience is $C/C_N = 10/12.5 = 0.80$. Note that both approaches satisfy the need for supply of $C = 10$ Gbps for loss of a single satellite. It also becomes apparent that *proliferation can be viewed as a special case of distribution,* as an initial constellation of five satellites could have been chosen, with a sixth added for proliferation, which is equivalent to the original example of distribution. But limiting the solution space to the proliferation case may result in a non-optimized solution, with more margin than required, usually at additional cost.

A closer look at uniform distribution versus proliferation over a threat range reveals that the special case of proliferation provides greater benefits at higher threat levels. Figure 6.6 shows the comparison between regions of advantage for uniform distribution and proliferation more clearly. This example is for a system with a total capability $C = 1$ distributed equally among eight elements. The desire is to maintain the full capability even when threats cause loss of

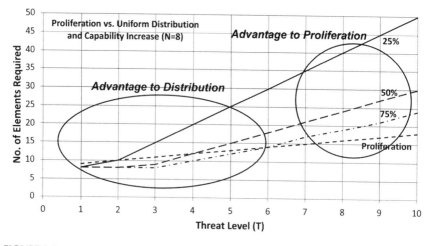

**FIGURE 6.6**
Regions of advantage for uniform distribution and proliferation.

elements (T = threat level = element loss). Clearly to design such a system, additional capability must be added, and then the new total must be more widely distributed. Three different amounts of added capability to be distributed among the elements are shown (25, 50, and 75 percent), and for proliferation. As the graph shows, for smaller amounts of added capability (e.g., 25 percent), the greater the level distribution (more elements) is required to sustain the threat and deliver the required resilience. As the threat level increases, the special case of proliferation becomes more advantageous in terms of minimizing the total number of elements required.

The detailed calculation for the graph in Figure 6.6 can be derived by considering the *minimum amount of additional capability that is required* given that a system with N elements is further distributed by adding M additional elements. This is important because excess capability has a cost, which should be minimized. That extra capability varies depending upon the degree of distribution selected. The larger the total number of elements (N + M), the smaller the additional capability required to meet a minimum capability value $C_{min}$ for a given threat level (T).

The relationship between the capability, threat level, and element count for any system can be expressed mathematically. For a system with an initial number of N elements providing a total mission capability $C_{min}$, the desire is to increase resilience by increasing the element count by M elements. For this to be effective, excess capability must be added to obtain a system with a new capability of $C_N$, which is to be minimized. The threat level to be mitigated is T, which represents the maximum number of elements that can be lost while maintaining the desired resilience. The relationship between the total number of elements (N + M) and T and the fraction of added capability ($C_N/C_{min}$) is expressed by Equation (6.1):

$$N + M \geq T\left[\frac{\beta}{\beta-1}\right] \tag{6.1}$$

$$\beta = \frac{C_N}{C_{min}}$$

where $\beta$ = Ratio of the new capability ($C_N$) to the minimum capability ($C_{min}$)

Consider an initial minimum system capability $C_{min}$ provided by N = 8 satellites, which is the minimum number required to meet system requirements, with a desire to limit the capability increase to 20 percent ($\beta$ = 1.2). To sustain a threat resulting in the loss of five satellites (T = 5), a total of N + M $\geq$ 30 satellites would be required. This case is shown parametrically in Figure 6.7. If the allowable excess margin can be increased to 40 percent, the total element count could be reduced to 18. The required element count increases with increasing threat level and decreasing excess capability. The graph also shows the effect of simply adding more of the same sized

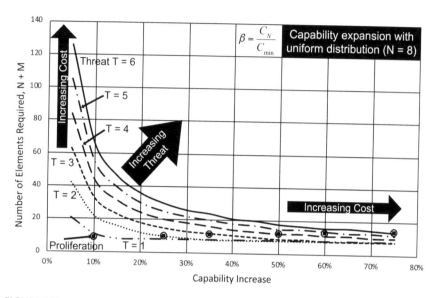

**FIGURE 6.7**
Required number of elements versus increase in excess capability and threat level.

element (proliferation), as opposed to adding capability and then optimally (and equally) distributing it. As the amount of required additional capability increases, the results of the two methods converge, as expected. Costs will increase with element count as well as the amount of added capacity to meet resilience goals, so there is a trade-off to obtain a minimum cost for a pro-scribed minimum system resilience and performance.

A family of curves showing the required number of additional elements ver-sus the required system-level resilience can also be generated (Figure 6.8) with a parametrically increasing threat level (e.g., number of lost elements due to successful attack). It becomes evident that the key relationship is between the required resilience, the number of uniform elements (N), and the threat level (T). Simple cases are limited in utility but are instructive to get a sense of the trade implications. The analytical equation can be inverted to provide a design equa-tion involving these three variables. In this case, $R_{RQ}$ is the *minimum required resilience* of the system (based on the required mission capability), and M is the *minimum* number of additional elements *that must be added* to a baseline system with N elements to achieve the minimum resilience for threat T.

The resulting Equation (6.2) is:

$$M \geq \frac{T}{1 - R_{RQ}} - N \tag{6.2}$$

In this case the baseline architecture has eight elements, and the threat level varies from 1 to 4. This clearly shows the need for a large number of elements

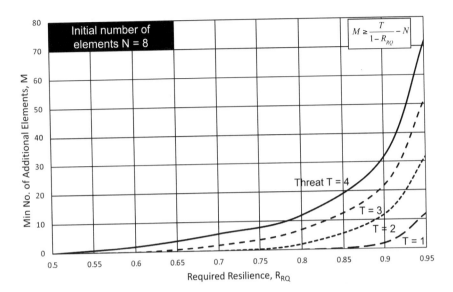

**FIGURE 6.8**
Minimum number of additional elements versus required resilience and threat level.

as the required resilience exceeds approximately 0.85. Here the designer selects the required resilience and the number of *additional* elements is calculated parametrically by threat level. For the eight-satellite baseline, a threat $T = 4$ yields a resilience of $4/8 = 0.5$. To raise the resilience to 0.90 requires adding 32 satellites for a total of 40. Note that in this example, no additional capability is added to the system; increased resilience is obtained simply through increasing the level of distribution by distributing the capability among a greater number of elements, given that some loss is tolerated.

While these relationships are useful and instructive, the cost part of the trade may overwhelm the rest of the trade variables. The costs of implementing a range of architectures is expected to vary considerably and while some of these solutions may be highly desirable from a resilience and performance point of view, the cost of implementation may be prohibitive, so it is essential to keep the total system cost in mind throughout the design phase, including the ground and user impacts; proliferation of the space layer will add complexity to the ground as well.

In particular, it is important to understand the driving constraint that limits the maximum number of elements as well as their complexity (capability per element). In the case of satellites, is it cost, availability of spectrum/orbital slots, mission planning and operations complexity, or something else? If reductions in satellite and/or launch costs can be achieved, perhaps there is no increased cost to distribution in that axis, skewing the trade in that direction.

In addition, each threat mitigation technique should be evaluated for its robustness to evolving threats. Some mitigations may have a short shelf life, limiting the time frame over which they are expected to be effective. This may effectively cause a pro-rating of the value of that mitigation approach, decreasing over time as the threat it is countering evolves.

### 6.2.4.1 Uniform versus Non-Uniform (Mixed) Architectures

Architectures may be comprised of identical, uniform elements, as shown in previous examples, or may include greater diversity in size and capabilities. A simple way of introducing diversity into a system is considering elements of different sizes, each with proportionally different capability levels. For the case of communications satellites, this can be represented by payloads of multiple sizes, each with a specific capacity. For the space segment the element sizing can include smaller payloads hosted as secondary missions on larger satellites or dedicated satellites in which the payload is the primary mission, up to much larger satellites providing a significantly greater capability. Assembling a system from these building blocks results in a more diverse and non-uniform architecture. This precipitates a key trade between increasing resilience through distribution versus diversification. A quick look at the impact of various types of distribution can provide insight into this trade.

Figure 6.9 shows a comparison of the resilience profiles for a number of uniformly and non-uniformly distributed architectures versus the threat level. Not surprisingly, as the number of elements is increased the resilience also increases independent of whether the distribution is uniform or non-uniform. But in comparing architectures with an equal number of elements,

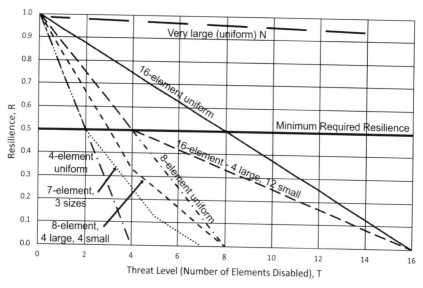

**FIGURE 6.9**
Resilience for uniform and non-uniform distributed architectures.

*the non-uniform architecture will always have lesser resilience over some portion of the threat range than equivalent uniform architectures.*

This conclusion is based on the worst-case assumption that the larger elements with greater capability are disabled first (as shown in Figure 6.9), causing a disproportionate loss relative to a uniform case in which all elements are equal. In the case of hostile actions, the worst-case assumption is that a sophisticated adversary has sufficient knowledge to target the more highly valued assets first. Even in the case of adverse conditions, putting too much capability into fewer elements presents the possibility of reduced resilience against a given threat than if that capability is more uniformly parsed across the system. Note that as the number of elements N is increased, the resilience across the entire threat range increases until, in the extreme, it approaches 1.

In this example, two different eight-element architectures are shown: one is uniformly distributed, while the other includes four large and four small elements. For the entire threat range (except the endpoints) the uniform case provides greatest resilience. The conclusion is that if resilience is valued above all else, the preference is to create more uniform architectures without any weak links, as these architectures do not provide an adversary with any local vulnerability resulting in a benefit to targeting some minimum set of elements. More elements are better and greater uniformity results in improved resilience.

Note that the ratio between the sizes of the element capability values also has an effect on the resilience. As the ratio between largest and smallest elements increases, the mid-range resilience decreases. As this ratio approaches 1, there is a convergence to the uniform profile. Figure 6.10

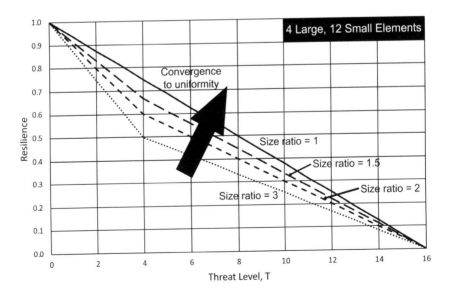

**FIGURE 6.10**
Convergence of non-uniform to uniform architectures.

shows a two-level architecture with four large elements and 12 smaller elements. The size ratio is varied parametrically from 3 to 1 to illustrate this convergence.

This does not imply that there aren't other ways to address this situation. This example makes simplifying assumptions, in particular that there are no avoidance or recovery features included in the system design. If other threat mitigation features can be affordably introduced into one or more of the elements, the resilience response can be changed to that shown in Figure 6.11, in which a non-uniform architecture can outperform a uniform case as the larger elements now have less chance of being disabled. This is another expression of the fundamental elemental resilience versus distribution trade that is discussed in detail in Section 6.2.5. Adding similar avoidance features to the baseline uniform architectures also provides similar results. Again, there may be other resilience, cost, and/or performance benefits to adopting a non-uniform architecture that outweigh any resilience improvement such as architectural flexibility and incremental scalability and these should also be included in the trade.

There can be great value in the incorporation of avoidance or recovery features to mitigate element capability loss. The greater the probability of avoidance, the less important it is to distribute for resilience, because any arbitrarily large element is so unlikely to be lost due to one or more threats. Moreover, different mitigations may be needed for differing threats, whereas distribution, as noted before, provides a more global benefit

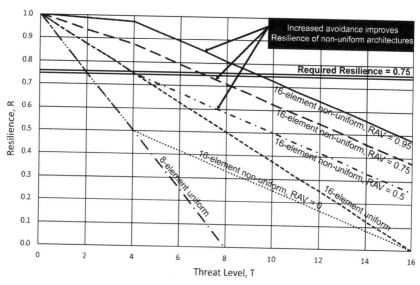

**FIGURE 6.11**
Resilience improvement for non-uniform architectures through avoidance.

addressing multiple threats by increasing the number of potential targets and decreasing the benefit of disabling any single one and thus increasing system-level robustness.

### 6.2.5 The Relationship Between Elemental Protection and Distribution

A fundamental trade in imparting resilience to a system is the extent to which individual system elements are protected (elemental resilience) and how much the system capability can be distributed to provide resilience through the architecture. These choices both depend on the capability delivered (as evidenced by the type of architecture required to deliver the capability) and the specific threats. This trade is generally illustrated in Figure 6.12. The greater the elemental protection, the less benefit there is to distribute solely for the purpose of increasing resilience, as the elements themselves become more highly resilient.

Note that the two techniques are not mutually exclusive, but rather can be combined to achieve the desired system-level resilience. This is more quantitatively illustrated for a point example shown in Figure 6.13. In this example, four of a system of N uniform elements are targeted. For simplicity, it is assumed that if the attack on an element is not avoided the element is fully disabled with zero robustness, recovery, or reconstitution at the element level. Only avoidance provides elemental resilience. Each of the targeted elements thus has a probability of survival, and it is necessary to calculate the expected number of surviving elements, resulting in the system-level resilience due to robustness through distribution. This system-level resilience is the expected residual capability, based on the number of surviving elements.

This calculation is made by using the probability of avoidance for the targeted elements in a binomial distribution equation to determine the expected value of the number of surviving elements, normalized by dividing by the

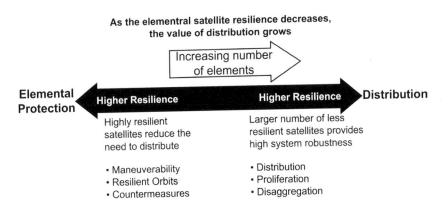

**FIGURE 6.12**
Balancing protection and distribution.

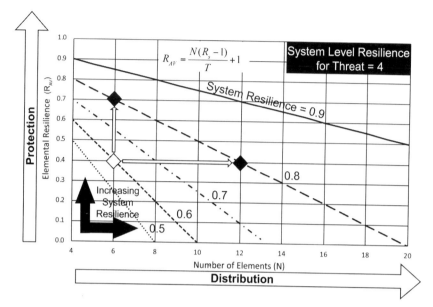

**FIGURE 6.13**
Balancing elemental protection and distribution to achieve resilience.

original number of elements to obtain a resilience value. It can be shown that the resulting equation describing the relationship between the required elemental resilience and the number of elements and threat level for a desired system-level resilience as shown in Figure 6.13 is expressed as Equation (6.3):

$$R_{AV} = \frac{N(R_S - 1)}{T} + 1 \qquad (6.3)$$

where N is the number of elements, T is the number of elements targeted, $R_s$ is the desired system-level resilience, and $R_{AV}$ is the probability of avoidance providing the elemental resilience. This equation is only valid for $N \geq T$, where in this example T = 4. There is also a minimum value for $R_s$ such that $R_{AV}$ becomes zero. The result of this calculation is a family of lines of constant system-level resilience. In Figure 6.13, the line for a constant system-level resilience of 0.8 is expressed as $R_{AV} = (N/4)(0.8 - 1) + 1 = -0.05N + 1$, for example.

This graph illustrates that there are multiple combinations of elemental protection ($R_{AV}$) and distribution (number of elements, N) that will produce the required system-level resilience. Figure 6.13 illustrates two ways of increasing the system-level resilience for a six-element system with elemental resilience of 0.4 from 0.6 to 0.8. The first approach is to raise the elemental resilience for each element from 0.4 to 0.7. The second approach is to further distribute the capability across 12 elements, each maintaining the original elemental resilience of 0.4. Thus, system-level resilience comes in two distinct flavors: protecting each element individually and distributing the system capability to achieve system-level robustness.

But there is still the matter of determining the optimal combination. Moving along the lines of constant system resilience imply that the requisite system designs also change and the system cost (and perhaps performance) can be expected to change with it. Figure 6.14 overlays a notional cost function on top of Figure 6.13 to further illustrate how cost may vary by both elemental protection and distribution. Elemental protection has a cost (including the development costs) and the higher the required resilience, the higher the cost. Likewise, it is clear that distribution also has a cost as a capability is apportioned across multiple elements (also illustrated in Figure 6.7). This extra cost is due to the additional satellites and the additional launch vehicles required, as well as any associated ground segment costs for managing a greater number of satellites or payloads.

In Figure 6.14 a maximum acceptable cost is chosen that determines the range of protection/distribution options that meet this constraint. In this example, achieving a minimum system-level resilience value of 0.8 can be achieved in multiple ways that will result in a cost below the target value. A system of six satellites each with an elemental resilience of 0.70 lies right at the target value, however, other points on the constant resilience line, including a system of 12 satellites with elemental resilience of 0.40, are firmly below the line. This example reinforces the need to be able to obtain costs for the available threat mitigation features and architectures in order to accurately assess the trade.

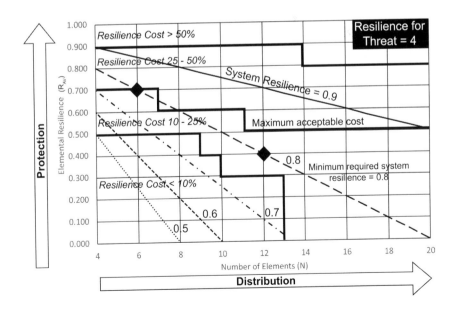

**FIGURE 6.14**
Optimizing resilience options by cost.

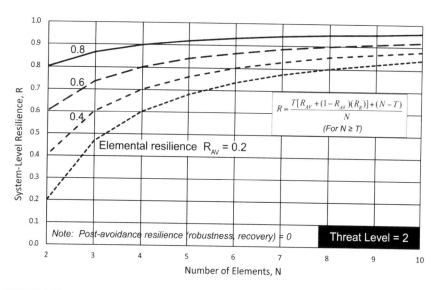

**FIGURE 6.15**
The effect of elemental resilience upon system resilience.

The value of protection and distribution may also vary depending on the threat and the mitigations available. Figure 6.15 illustrates this by showing the system-level resilience R versus the number of elements N and the elemental resilience $R_{AV}$ for a specific threat level (in this case a threat level of 2, with a post-avoidance resilience $R_R$ of 0). Equation (6.4) is derived from the equation in Figure 5.2, with weighting of capability for N elements, T of which are threatened:

$$R = \frac{T\left[R_{AV} + (1 - R_{AV})R_R\right] + (N - T)}{N}$$

(6.4)

This graph further demonstrates that as the elemental resilience is increased (increased avoidance), the benefits of distribution fade. Furthermore, as the number of elements increases far beyond the threat level, the incremental system resilience gains also narrow, providing diminishing returns. The use of simple calculations such as this can aid the designer in finding the regions of maximum leverage in the design of the system.

## 6.2.6 Cost versus Resilience Trades as Function of Distribution

At this point it becomes clear that when considering distribution, the true trade between cost and resilience is one of where to best invest resources to obtain the highest probability of success. Figure 6.16 further illustrates the trade, using the space segment as an example.

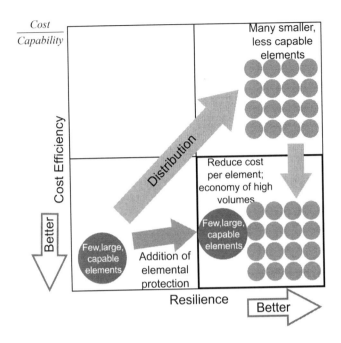

**FIGURE 6.16**
The economics of the distribution trade.

The motivation to retain systems with large, highly capable elements (such as satellites) is largely based on fully leveraging the cost advantages of placing large, capable, yet expensive satellites on large, capable, yet expensive launch vehicles. This results (at least theoretically) in high cost efficiencies for which the cost metric is often expressed as cost per unit of capability. In practice this may not be true if the non-recurring development costs become prohibitive as can happen for leading-edge, complex systems. For a SATCOM system, this cost metric could be expressed in terms of cost per unit of bandwidth. Aggregating into large satellites delivering a high capacity tends to reduce this value due to the high fixed cost of launch, something that is highly desirable.

For an alternative architecture with many smaller satellites providing the same total capability the cost is expected to be greater per unit of capability. Using today's practices, there are more launches, more overhead due to cost of satellite buses, and so forth. But the advantage is higher system-level resilience.

There are two paths to achieving the ultimate goal of both high cost efficiency *and* high resilience. The first path is to add elemental threat mitigation features to the space segment to enable retention of the few large satellites while increasing the resilience. This, of course, requires some additional investment, so one expected consequence would be to reduce the net cost efficiency to some degree. The second path is to apply new paradigms to the many-satellite constellation to reduce the average cost of launch per satellite, and the average cost of each satellite itself to improve the overall cost efficiency, with small or no loss of system resilience. This may reduce to a battle between the economies of high fixed launch costs and high-volume production.

## 6.2.7 Diversifying Distributed Architectures

The U.S. Office of the Secretary of Defense (OSD) identified diversification as one of the six key resilience sub-elements, and it brings with it many benefits, even beyond resilience itself. Diversification provides alternate ways of delivering capability, such that certain threats may not affect all of these paths equally, if at all. Diversification can also provide inherent protection against systemic threats that could affect all of one type of system element. Diversity shares some similarity to the use of redundancy to increase reliability. But whereas redundant paths often use identical means, diverse paths provide capability in substantially different ways, with differing attributes. This difference in delivery method dilutes the effectiveness of a threat across multiple, diverse delivery paths.

Diversification may be achieved through the use of multiple frequency bands and waveforms, commercial and purpose-built satellites, and/or use of satellites in multiple orbits. Diversification is most effective when it provides resilience to different parts of the system to different threats. In the case of hostile actions this forces an adversary to invest to diversify the threat assets, which may lessen the overall threat effectiveness given limited resources.

An often-cited example is a MILSATCOM system that is a hybrid of both government-owned satellites and commercially leased satellite services. Government and commercial satellites use different frequency bands to transmit and receive signals and thus the use of both concurrently can limit an adversary's effectiveness in denying service using jammers operating in a single frequency band. They also employ different types of antenna coverage patterns and may use different cryptographic technologies to protect command links. In addition, there may be additional ramifications to disruption of commercial satellite services that may provide an amount of deterrence, complicating the adversary's decision process.

Of the many methods of distribution, some contain more or less diversity than others, and have varying effects on resilience. By considering these differences the designer can trade resilience provided by distribution versus the value of a more diverse system design. As with any such trade, the associated costs will also vary according to the level and type of diversification. The cost of diversification will be examined in Chapter 7.

## 6.2.8 Disaggregating a System Architecture

Disaggregation is the decoupling of missions or key functions within a system architecture. Disaggregation more often provides cost and/or performance benefits but can also be employed to increase resilience. Many space systems support multiple missions, often through the use of general purpose elements that can meet the diverse requirements without significant differentiation of the implementation. A wideband transponded satellite system can provide high bandwidth communications for a number of different missions without specialized hardware on board each satellite customized to individual missions. Sometimes this is not true, and when a satellite supports multiple missions, unique payloads must coexist on a single platform. This can lead to compromises in performance and resilience, although this approach is often intended to minimize cost.

If the mission payloads are placed on separate platforms, the cost increases, but the performance for each mission can be increased by tailoring the individual satellites. This provides the greatest benefit for missions in which the respective requirements differ significantly. For example, a tactical system may require high capacity in a small geographic area in one part of the Earth, while a strategic system may require less bandwidth, but global coverage. Aligning these requirement sets requires compromises if both are housed on a single satellite.

Likewise, resilience can also be increased through disaggregation. Threats are often uniquely target-specific missions. Placing multiple mission payloads on a single satellite can result in an aggregation of threats, requiring more complex threat mitigation for these high-value assets. Additionally, the most stringent requirements for each of the missions can be required of all of the satellite payloads, increasing the cost to mitigate the sum of the threats. Satellite nuclear survivability is an example of this. A strategic satellite hosting tactical payloads may pass along the requirement to be nuclear survivable to all payloads, not just the strategic one.

Disaggregating missions such that each design can be tailored to optimize threat mitigation for mission can not only reduce the mitigation costs, but result in increased resilience and decreased satellite complexity due to the decoupling of threat effects.

## 6.3 Evaluating Resilience for Multiple Missions and Threats

Chapters 2 and 4 described scenarios with multiple threats to a system as well as ways to calculate the system's resilience. When a system services multiple missions then additional consideration must be given to the impact of each threat on each of the missions in question.

### 6.3.1 Systems Supporting Multiple Services or Missions

Many space systems provide simultaneous support for more than one service or mission. In fact, the less specialized the capability being delivered, the more likely the system meets multiple needs. In these cases, a single threat

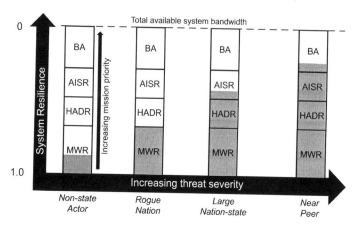

**FIGURE 6.17**
Resilience implications for systems supporting multiple missions.

can have varying impacts upon the individual services or missions that are being supported. This is because the minimum required capability (and thus required resilience) likely varies by the service or mission.

Figure 6.17 illustrates this case for a MILSATCOM system, with the system delivering an aggregate capability of regional communications bandwidth that supports four unique missions: Battlespace Awareness (BA); Airborne Intelligence, Surveillance, and Reconnaissance (AISR); Humanitarian Aid and Disaster Relief (HADR); and Morale, Welfare, and Recreation (MWR). The total system bandwidth is apportioned among these four missions in proportion to the minimum requirement for each. In this example, MWR consumes the most bandwidth, followed by BA, AISR, and HADR. Each mission is also assigned a priority, independent of the amount of bandwidth that each uses. Here the BA mission has the highest priority, though it is not the greatest consumer of resources. Typically, for military systems the mission's priority is related to the conflict level. Missions crucial to a conflict receive higher priority, particularly in contested environments.

In Figure 6.17 the impact of an increasing threat severity is shown as the threat impact increases from left to right, based on the level of adversary and their capabilities to create greater levels of disruption. As the realized threat severity increases, the usable bandwidth decreases. In Figure 6.17, the denied resource is illustrated by the shaded area. As the severity increases, the available capacity decreases. The remaining capacity is reserved for the higher-priority missions, with each mission being sequentially preempted in order of priority as the threat's impact increases.

In this example, the shared resource (capacity) is incrementally denied with increasingly more severe threats and the system operators must reallocate among missions to ensure that those of highest priority are affected the least. In this way the impact of a single threat can be different for each mission.

The MWR mission requires a significant amount of the total (allocated) capability and requires a *high* system-level resilience to meet its mission requirements. In a contested environment at a high conflict level that mission has the lowest priority and thus under those conditions the mission may not meet its minimum requirement because other missions have higher priority and are allocated the remaining usable resource.

As the threat increases the next mission to be impacted would be the HADR mission. A system resilience of 0.65 would still permit that mission to meet its mission requirements. As the threat severity increases, such as a threat level presented by a large nation-state level adversary, a system resilience of as little as 0.5 may be sufficient to meet the AISR and BA mission requirements. Beyond that, even those missions will be adversely impacted.

This example illustrates that if the capability can be flexibly reassigned to users, reallocation of remaining capability can be a means of recovery for one more of the missions (at the expense of others). This example is instructive to show that any design for resilience activity should consider the requirements for all of the missions serviced, not just one, and to understand the relative priorities.

The ability to reallocate bandwidth among missions is also a crucial part of the design for resilience. If the system is not designed to be flexible enough to enable easy and responsive reallocation of resources, then the resilience may suffer for multiple missions. This emphasizes how design choices can directly affect resilience even apart from specific threat mitigation features.

## 6.3.2 Comparing Architectures Across Multiple Threats

In Chapter 5, Section 5.1.3, a method for combining threat effects into a single value was presented. This approach requires estimates of threat probabilities to weight the individual threat effects. This detaches the final resilience value from an expected value interpretation and its value is less obvious other than purely as a comparative metric. The probability of a threat occurring may also be difficult to estimate. An alternative is to calculate the individual resilience values across all threats and tabulate them for a more global comparison among all candidate architectures.

Though there may be a desire to represent the system-level resilience over all threats and missions using a single number for evaluation purposes, this is likely impractical and is opaque to the designer as well. More likely the resilience analysis results in a table similar to Figure 6.18. Though this may seem more complicated, having an array of values can also provide insights in comparing candidate architectures. In the provided example, it is clear that all of the candidate architectures are insensitive to threat scenario 3, but that there is much greater variation for threat scenarios 4 and 5. Detecting these correlations can help guide design trades, particularly when the threats are prioritized in some manner.

| Architecture | Threat Scenario | | | | |
|---|---|---|---|---|---|
| | 1 | 2 | 3 | 4 | 5 |
| A | 0.80 | 0.66 | 0.99 | 0.23 | 0.92 |
| B | 0.75 | 0.77 | 0.98 | 0.33 | 0.96 |
| C | 0.78 | 0.70 | 0.99 | 0.67 | 0.74 |
| D | 0.60 | 0.75 | 0.97 | 0.19 | 0.64 |

**FIGURE 6.18**
Summary resilience table for multiple architectures and threats.

## 6.3.3 Sequential or Recurring Threats

The simple examples discussed so far have considered a single threat event. But there may be threats that are defined by multiple, sequential events or of some repetitive nature, that occur more than once during the system lifetime or time frame of interest. In this case, resilience must be evaluated across the entire event period, taking into account the cumulative impact to the system due to all of the anticipated events. Section 5.1.4 in Chapter 5 introduced the concept of multiple threats and the technique of superposition to evaluate the consolidated system response. That approach is more desirable when the multiple events occur closely in time. For events that may be separated by longer periods of time it can be useful to continue to view them as truly separate events and treat them that way analytically.

Figure 6.19 illustrates a simple case of a system impacted by two sequential events. The first event occurs and the system's resilience to the threat is 0.875, irreversibly losing 12.5 percent of its initial capability. This is calculated

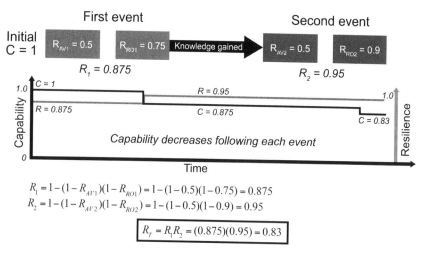

$$R_1 = 1-(1-R_{AV1})(1-R_{RO1}) = 1-(1-0.5)(1-0.75) = 0.875$$
$$R_2 = 1-(1-R_{AV2})(1-R_{RO2}) = 1-(1-0.5)(1-0.9) = 0.95$$

$$\boxed{R_T = R_1 R_2 = (0.875)(0.95) = 0.83}$$

**FIGURE 6.19**
Calculation of resilience for sequential or repetitive events.

based on an avoidance value of 0.5 and a robustness of 0.75. Both recovery and reconstitution are assumed to be zero. If a second event occurs as the result of a second identical threat, there will be further degradation to the capability.

In Figure 6.19, there is time to perform a threat analysis following the first event, which yields knowledge leading to a reconfiguration of the system resulting in an increased robustness of 0.9 to the threat that has just been sustained. The avoidance probability of 0.5 is unchanged. This results in an increased resilience of 0.95, although referenced to a reduced capability level. When the second event occurs (a repeat of the first event), the capability of the system is further reduced by 5 percent of the 0.875 value to 0.83 (relative to the original capability of 1). Subsequent learning may lead to further increases in the resilience to future threats, but unless it becomes 1, capability loss is expected to continue incrementally if additional threat events are encountered. Note that the overall system resilience, $R_T$, is the product of the resilience values for the two events.

This method may be repeated for as many events as required. In the case of a repetitive event that can be predicted to occur at some regular (or irregular) interval, if the capability loss for each event is known, the time to minimum required capability may be calculated which may be the useful life of the system assuming that there are no other remedies.

## 6.4 Including Threat Effectiveness

One of the more difficult resilience parameters to estimate is the effectiveness of a threat. Worst-case estimates of $P_k$ will tend toward 1 but this may overestimate the effectiveness resulting in overdesign and unwarranted cost. As a result, it is useful to evaluate the impact of threat effectiveness parametrically to understand the sensitivity of the system's resilience to that parameter. As the effectiveness varies, so will the resilience to that threat.

Consider a space segment consisting of 10 elements of uniform capability, each of which is targeted by a threat. The threat effectiveness for each of these threats is a constant $P_k$. What is the *probability of N or more elements surviving* the attack as $P_k$ varies, and assuming no inherent threat mitigation is present in the system design? Using the general form of the cumulative binomial distribution, Equation (6.5), over a range of 10 satellites results in the probability profiles shown in Figure 6.20.

$$P(X \le x) = \sum_{x=0}^{N} \frac{n!}{(n-x)!x!} p^x q^{n-x} \tag{6.5}$$

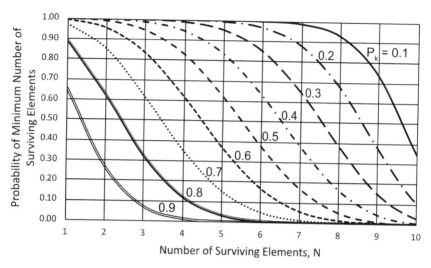

**FIGURE 6.20**
Probability of number of surviving elements versus $P_k$.

The binomial distribution returns the probabilities for the loss of 1 to N satellites, given the probability of a successful attack on each ($P_k$). As expected, as $P_k$ increases, the probability of any given number of survivors decreases. If there are no other resilience terms this value could be the resilience of the system. Note that since the total number of system elements equals the number of elements attacked (N = T), T is not a variable here. Referring to Figure 6.20, for a $P_k$ of 0.8, or 80 percent, the probability of 5 or more satellites surviving is about 4 percent. If the system resilience must be 0.5 or greater, requiring 5 of 10 satellites to survive such a threat event, a much higher probability would be desired. This means that $P_k$ must be 0.4 or less to result in a greater than 80 percent probability that 5 or more satellites would survive, a low effectiveness value for a credible threat. This confirms the assertion that for highly effective threats, threat mitigation features are required to produce high resilience systems.

## 6.5 Other Statistically Relevant Considerations of Resilience Calculations

While the resilience calculation yields an expected value for probabilistic cases, this is not the same as establishing a confidence level based on the variance of the variable. Consider the previous example (Figure 6.20), with a $P_k$ of 0.5. The calculated *resilience* value for this case can be found to be 0.50,

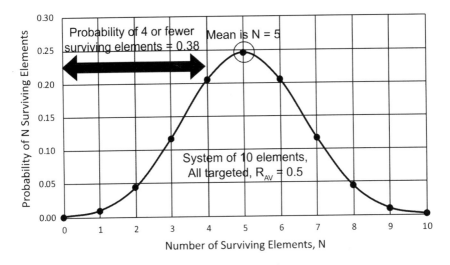

**FIGURE 6.21**
Probability distribution of N surviving elements versus N ($P_k = 0.5$).

reflecting an expectation that 5 of the 10 satellites will survive. But the *probability that 5 or more will survive* is 62.3 percent (Figure 6.21). If the minimum required resilience is 0.5, there is a 62 percent probability of that value actually being achieved. In this way, resilience is a mean value, and thus does not include the variance of the random variable for probabilistic results. However, this variance is the same when applied equally to a number of candidate designs, and so should not bias comparative results.

Note that given the potential inaccuracies in threat determination and other factors, combined with the general need to make comparative measures, this measure may not be that meaningful to the designer. If the value is low enough to give concern, then it may cause the consideration of adding additional system margin to assure uninterrupted service.

## 6.6 Resilience Design and Analysis Tools

Currently there are no publicly available modeling, simulation, and analysis tools for automating the design for resilience process. In time this may change if the demand is great enough. Nevertheless, it is instructive to consider the key functions that are required of such tools in the case where a homebrew solution is desired.

The tool should be capable of importing the system model from an existing design tool. This may be at a high level, it could be using an established model-based systems engineering (MBSE) tool, but it must contain the information required to describe the system for resilience design and analysis purposes. If this is not possible, the tool itself should have an elemental database and the means to construct the system model from within.

The resilience tool should also either contain or be able to access a threat database that provides the relevant threat information. The final piece is a database of threat mitigation features that may be implemented in the system design, and their key features and performance. Ideally, the cost of implementing these features would also be included in the database. Specific mitigation design features can then be imported into the system model as the design is iterated and evaluated. All of the detailed lower-level analysis tools may not be included in the design for resilience tool, but the analysis outputs should be available through standard data formats and interfaces to allow lower-level tools to be linked to the resilience tool.

# 7

## Applying Resilient Design Techniques

The tools provided in the previous two chapters form the basis for performing resilient space system design. These resilient design techniques and trades are applied to the design, development, and optimization of resilient systems. Candidate system architectures are developed and traded against one another, and specific threat mitigation features are imparted to the design to achieve the desired result. This process is iterative, with the designer continually revisiting the projected cost, performance, and resilience and balancing each according to the prioritization of the requirements. This chapter presents a number of simple examples to illustrate this design process and the application of the available resilience tools and techniques that were previously introduced.

The examples are based primarily upon satellite communications (SATCOM) systems because such systems are most useful for illustrating the application of the design process. The design of other types of space systems will follow a similar process.

## 7.1 Creating a Resilient Space Architecture

Developing resilient space system architectures is a process of interpreting the system requirements, allocating and assessing the required resilience, trading the available options, and then implementing the necessary mitigation approaches to achieve an optimal system design. This is an iterative design process with the number of iterations increasing with the complexity of the system. The implementation cost is evaluated with each new iteration as well as the system performance to ensure that all key parameters are balanced in accordance with the established priority.

This chapter presents simplified scenarios to illustrate increasingly complex space system designs. Scenarios with single and multiple threats are examined to highlight the design choices necessary to increase the system resilience to the required level. Threats against both space and ground segments are also considered. Notional cost values are included to represent the entire trade space. Many of the resilience mitigation techniques from previous chapters are leveraged, including increasing elemental resilience, creating more distributed architectures, using diversification, and increasing threat avoidance.

### 7.1.1 Multilayered Architectures

In many of the examples presented in this chapter, the trades involve how capability is apportioned among different kinds of system elements or services. One useful approach is to view a system segment as a number of separate layers to help visualize the methods of delivering a capability within the larger system.

Space system architectures with groupings of similar satellites are examples of "multilayered architectures." This means that the satellites (and sometimes the ground segment) can be represented as a number of layers, with each layer defined by one or more specific properties possessed by the satellites or services within the group. This grouping into a layer can be based on the specific type of satellite or amount and/or type of capability it delivers, its size, its mission and users, or some other property that distinguishes it from those in the other layers.

An example is the current U.S. military satellite communications (MILSATCOM) architecture (Figure 7.1). A small group of strategic satellites provide nuclear-survivable assured communications globally and represent a highly resilient layer to multiple threats. A second layer can be described that includes satellites that provide tactical narrowband and wideband services for a much larger number of users with less resilience to RF and nuclear threats. Finally, a third layer encompasses leased services provided by commercial satellites servicing even more users with fewer military-unique features. Each of these layers is distinguished from the others by the operating frequency bands, the services provided and/or missions supported, the number and types of users and their terminals, and the performance (such as capacity) and their resilience to threats such as jamming.

Alternate views of the same system design are also possible. In the MILSATCOM example, the space segment can also be partitioned according

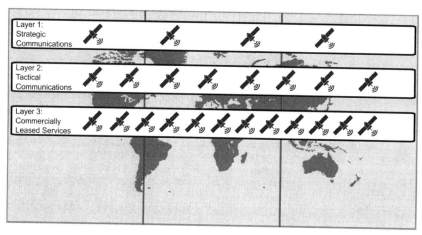

**FIGURE 7.1**
U.S. MILSATCOM layered architecture view.

to its functionality and is sometimes represented as such. In the past, these categories were nuclear strategic, wideband communications, and narrowband communications. In this view, both the wideband and narrowband groups can contain a combination of government-owned, purpose-built satellites as well as commercial services. This shows that the choice of view is dependent on the design approach and the system requirements and top-level architecture. A more recent view generated by the U.S. Department of Defense (DoD) is grouping systems by their threat environment: *nuclear, contested,* and *benign.* However these layers or groupings are defined, the goal is to make use of them in some optimal combination to obtain the desired result.

In a communications system the users must be able to access services from one or more layers. Any given user may or may not be compatible with multiple layers, depending on the number and type of user terminals available at the point of use. Certain wideband military terminals are "dual-band" and can access both government-owned and commercial satellites using different frequencies, protocols, and waveforms, providing flexibility. The user terminal requirements are often heavily influenced by the platforms that they are installed on and can become system-level design drivers.

A layered architecture can also be viewed based on the resilience of its elements. Resilient architectures can be formed using a layered approach, with each layer providing a different level of resilience due to inclusion of different threat mitigation features. These features include satellite security features (cyber threat) and anti-jam features (electronic threat). Through its resilience features, each layer can be tailored for a particular subset of services, missions, or users. In particular, more resilient layers might be dedicated to higher-priority or mission-critical traffic.

Certain government space architectures also can be represented in a slightly different layered view, with a government-owned "core" layer of purpose-built satellites, and a commercial services layer that provides "augmentation." This concept of a core layer and augmentation layers is an increasingly preferred way of thinking about how to meet government military and civil needs through a diversified approach. The core layer often provides the highest level of assured access, sovereignty, and control while the augmentation layer provides services that are less assured or resilient but are often complementary in some way and can be accessed quickly to satisfy an urgent need or surge requirement. In terms of resilience, this can be viewed as a multilayer architecture in which there is a high resilience layer and one or more layers with lower resilience (Figure 7.2). The criteria for separating the satellites might be obvious, as in this case where there is a significant difference between government-owned military satellites and their commercial counterparts, but the true differentiator is based on the resilience to one or more threats. Each of these layers provides both benefits and drawbacks for the user.

In this example the core layer is more resilient to electronic threats which is the priority. This does not imply the relative amounts of capacity assigned to each layer, but simply that the core layer provides sufficient capacity with

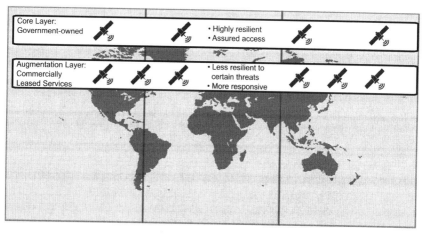

**FIGURE 7.2**
Core and augmentation layered architecture.

the required anti-jam protection. In fact, the augmentation layer could be providing the bulk of the total system capacity, but it might not be assured when confronted by certain threats. This layer can also be *more* resilient to a physical threat due to a greater number of satellites providing robustness through distribution of its capacity.

A layered architecture could also be formed by grouping satellites that provide identical services or capabilities but with different sizes and/or performance to meet the total system-level requirements. Figure 7.3 shows an example of a hybrid or mixed architecture for communications services. This architecture has two layers, one made up of five larger satellites to provide

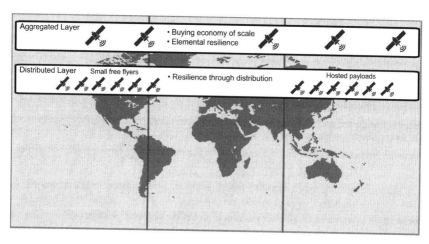

**FIGURE 7.3**
Hybrid resilient architecture.

cost-effective bulk capability, taking advantage of aggregated capacity on large platforms. The second layer is a highly distributed layer of smaller, free-flyer satellites and hosted payloads that make up the remainder of the capability. This second layer could provide somewhat lower performance in return for heightened system-level robustness due to the distributed nature of the layer. For a communications system, this might mean supporting lower data rates; for an imaging system, it might mean lower resolution of the imagery products.

## 7.2 Increasing Resilience Through Distribution

Using increased distribution of a capability is one way to increase system-level resilience. This scenario demonstrates how system requirements can be satisfied through changes to the architecture and thus achieve the required performance and resilience.

Consider the following requirements for a global SATCOM system:

1. The system is to provide communications services to users globally in the mid-latitudes (65°N to 65°S).
2. The total global capacity is to be equally distributed around the globe.
3. The operators can tolerate up to a 20 percent loss in capacity, meaning both the regional and global minimum system-level resilience is $R_{RQ} = 0.8$.
4. Each satellite must be able to contact at least two ground stations at all times.
5. A physical threat can result in the complete loss of one satellite; if a threat is effectively executed, the satellite is completely disabled without recovery.
6. An outage below the 80 percent level for more than 48 hours is unacceptable.

The simplest solution to satisfy this set of performance requirements is to establish a three-satellite geostationary equatorial orbit (GEO) constellation of identical satellites spaced equally around the equatorial arc with four ground stations located such that each is in the field of view of a pair of the satellites (Fig. 7.4). In this case the mid-latitude requirements are fully satisfied. As all three satellites are identical, and assuming relatively equal demand, each carries one-third of the global user traffic. Likewise, the ground stations also share the traffic from the satellites more or less equally. From a performance and affordability point of view, this is a straightforward and cost-effective solution so long as each satellite can provide the required capacity.

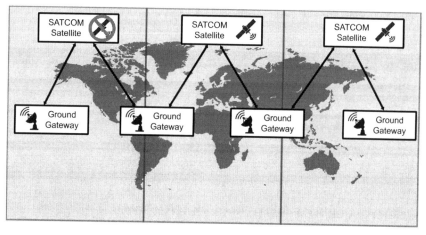

**FIGURE 7.4**
Three-satellite GEO SATCOM system provides global coverage.

But when considering resilience to the stated physical threat, this proves to be insufficient. The loss of any single satellite causes the loss of all services to one-third of the globe. In the past, the most likely cause of such a loss was an on-board failure. Satellite manufacturers responded by designing satellites to be very highly reliable, with many on-board features to overcome a component failure, including a high degree of on-board redundancy. This also allows for graceful degradation, depending on the failed part. But the threats that cause the loss of an entire satellite are external, such as orbital debris, and these reliability features have little value in increasing the resilience of the satellite. If an external threat disables a satellite in this scenario, the system-level resilience drops to 0.67, the resilience within one of the three global coverage areas is *zero*, and neither of the resilience requirements are met with the loss of a single satellite, with one-third of the Earth left unserved.

This implies that mitigation approaches are required, if available. Recalling Figure 4.1, the first mitigation approach to consider is some form of avoidance to provide elemental resilience and protect the satellites. This approach does not require any adjustment to the original baseline architecture. Instead, the resilience of each satellite must be increased from zero to at least 0.8. This assumes that the nature of the physical threat is well known enough to devise an effective mitigation providing the avoidance. If the robustness is zero, then the satellite is fully disabled if the threat is not avoided. Mathematically this meets the resilience requirement, but for a system with so few elements this solution could be considered insufficient by the operator if the threat is credible, as there is still a projected 20 percent chance that the satellite does not survive if the threat is directed at it, losing service to one-third of the Earth. This is an example where simply accepting resilience as the expected value of retained capability might not be enough to satisfy the operator.

Another solution is to attempt to build a more robust satellite, such that if impacted by the threat less than 20 percent of the capacity is lost. In practice, for a physical threat in particular, this is not credible due to the relative fragility of satellites. Likewise, recovery following a severe physical impact is unlikely given today's technologies. Reconstitution, through the launch of a replacement satellite for example, is possible but could be financially impractical as it would require that the operator have a replacement satellite ready to launch, including launch vehicle, to meet the 48-hour-outage requirement. This is likely an unaffordable solution, requiring an operator to invest heavily to purchase a spare satellite and launch vehicle that might not be required. In addition, launching within 48 hours is impractical today. So, avoidance is likely the most effective method of providing elemental resilience.

Increasing the resilience of individual satellites is not the only option, as architectural options can be employed as well. If the threat of concern is the loss of one satellite, then subdividing the communications capacity among a greater number of smaller satellites can accomplish the desired result. For the global requirement of a system-level resilience of 0.8, the minimum number of satellites required can be found to be 5 from the graph in Figure 6.3. Spacing these satellites uniformly can provide a minimum of two satellites servicing each coverage area that is one-fifth of the globe (Figure 7.5). Each of the original three satellites carried one-third of the global capacity, but if these smaller satellites remain identical then the five new satellites each now must only carry one-fifth of the total. If two satellites share a regional coverage, each can contribute half of its capacity to one of two overlapping coverage areas, each of which is now one-fifth of the global coverage area, requiring a fifth of the global capacity. Each of the two satellites can thus contribute $(1/2)(1/5) = 10$ percent of the global capacity, to meet the 20 percent regional need.

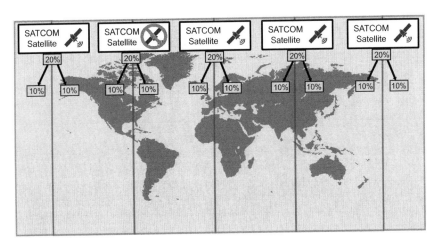

**FIGURE 7.5**
Five-satellite SATCOM system improves global resilience.

For this architecture, if one of the two satellites serving a region is lost the remaining capacity drops to 10 percent for a resilience of 0.5 which is less than the 80 percent regional requirement. So, this solution only meets the global resilience requirement, and an even greater level of distribution is required to meet the regional requirement.

To meet the regional resilience need, at least five satellites are required to be able to address any designated coverage area, such that the loss of one results in 80 percent of the capacity remaining for that region. Using GEO satellites, each with a view of approximately one-third of the Earth, or 120° of longitude, five satellites per region could satisfy the resilience requirement, placed with a spacing of 24° of longitude in the arc. The resulting ideal space-layer architecture would then be made up of 3 × 5 = 15 satellites. Depending on the frequency band, obtaining the required spectrum allocations might be problematic for a large number of GEO satellites, but this illustrates the architectural trade. Loss of any one satellite results in a loss of one-fifteenth of the total global capacity, resulting in a resilience of 93 percent, far surpassing the requirement of 80 percent.

The cost of this alternative could be prohibitive. This space architecture requires the purchase of 15 smaller satellites (Figure 7.6), each with approximately one-fifteenth the capacity of the global requirement, and the launch vehicles to place them into a GEO orbit. If the satellites are small enough, multiple satellites could be launched on the same launch vehicle, reducing the launch cost per satellite. But it is more likely that the cost of this fully compliant distributed system is more expensive than the original three-satellite system to counter the threat of the loss of a single satellite in order to meet global and regional resilience requirements. However, as the cost per satellite continues to decrease, this may not be true in the future.

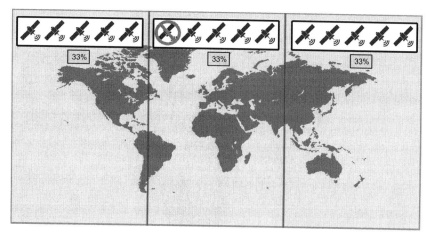

**FIGURE 7.6**
Fifteen-satellite SATCOM system meets global and regional resilience requirements.

Other options might be available to reduce costs. For example, if the communications payloads are small enough, some of them can be placed as secondary, hosted payloads on other GEO satellites, reducing the satellite and launch vehicle costs. Distribution can bring with it other costs, including the increased cost of the ground segment as now more antennas and data paths must be installed, and more geographic sites might also be required to support the increased number of satellites, unless some of the apertures can be time shared. This simple example illustrates the many factors that must be considered when trading cost, performance, and resilience.

Now consider a second threat: the loss of access to a single ground gateway. Assume that each satellite is in view of two gateways, as shown in Figure 7.7. If each gateway is capable of receiving a satellite's full capacity, then the second gateway in view is a secondary backup site and used only in case of the loss of the primary gateway. In this case, for any given satellite, the loss of a primary gateway can cause minimal interruption of service during the recovery time while the satellite and secondary gateway are reconfigured to restore service. If the satellite's traffic is split evenly and each gateway is only capable of receiving half of the satellite's capacity, then one-half of that satellite's capacity is lost and, for the original three-satellite case, one-sixth of the total global capacity is lost: $(1/3)(1/2) = 1/6$. Since $5/6 = 83$ percent of global capacity is retained, the global resilience requirement is still met. Half of the satellite's regional capacity is lost, resulting in a regional resilience of 0.5, below the 0.8 required, and thus again this is not a fully compliant solution.

Remedies to this are similar to the mitigation for loss of satellites, with additional ground sites providing greater distribution of ground-based capability to more robustly service the satellites. The number required depends

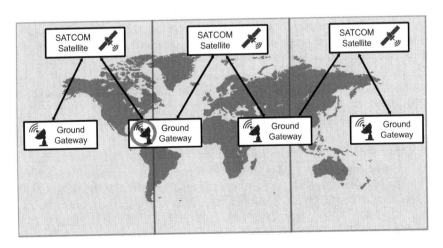

**FIGURE 7.7**
SATCOM system suffering loss of a single gateway.

on the system configuration and the way that the traffic is routed from satellite to ground site (and potentially beyond). In this manner, resilience solutions can be similarly implemented for all system segments.

## 7.3 Increasing Resilience Through Diversification

An example of a multilayered system architecture that achieves resilience through diversification is a purpose-built system that is augmented through the addition of commercial services. Diversification is the use of multiple means of delivering a capability that are different enough as to not all be affected by one or more specific threats. An example of this is the use of commercial SATCOM services to augment government-owned military systems. Military systems use frequency bands exclusively allocated for government use, including parts of UHF, X-band, and military Ka-band. Commercial systems use different frequency bands, including C-band, K-band, and commercial Ka-band. Jammers are most often built to operate on a specific frequency band, so the ability to use both commercial and military frequencies can increase the resilience of a system to a particular threat, such as an X-band jammer.

As with other mitigation approaches, there is usually an added cost for diversifying, and there can also be impacts to the resilience of the system to other threats. These factors must all be balanced when performing the system design. To illustrate the design considerations for a diversified architecture, consider a space system with four satellites providing 1.8 GHz of bandwidth each to a region of the Earth using military Ka-band frequencies. An electronic threat is posed by an adversary capable of jamming all four satellites simultaneously. The effect of a single jammer is to cause an effective loss of 60 percent of a satellite's capacity, so the elemental resilience is 0.4. If all four satellites are targeted, the system-level resilience is also 0.4.

A more diversified architecture is desired to counter this threat using a combination of both government-owned satellites and commercial services. This architecture is shown in Figure 7.8, in which one of the four government satellites has been replaced with communications services provided by a combination of eight commercial fixed satellite services (FSS) satellites, collectively providing 25 percent of the total system capacity using commercial C-band transponders. The Ka-band jammers do not affect the commercial communications links, meaning their resilience to that threat is 1. In the worst case, if the Ka-band jammers target the three government satellites, the system-level effect of the jammers has been reduced and the total system-level resilience is calculated as follows:

$$R = (0.75)(0.4) + (0.25)(1) = \mathbf{0.55}$$

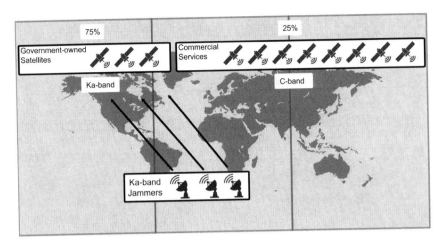

**FIGURE 7.8**
Diversified SATCOM architecture through addition of commercial services.

The use of a more diversified architecture has increased the resilience by 0.15, or 15 percent.

The cost of this approach must now be calculated. The cost trade is accomplished by comparing the expenditures for launching one military satellite providing approximately 1.8 GHz of bandwidth versus the cost of leasing a number of commercial transponders providing the equivalent bandwidth over the lifetime of the government system (15 years). This equivalency occurs when Equation (7.1) is true:

$$\text{Cost of the satellite and launch} = \#\,\text{TPEs} \times \$/\text{TPE}/\text{year} \times 15 \text{ years} \quad (7.1)$$

where a TPE is a transponder equivalent bandwidth of 36 MHz.

The number of TPEs required to replace the 1.8 GHz of owned bandwidth can be found by employing Equation (7.2):

$$Number\ of\ TPEs = \left\lceil \frac{1800\ MHz}{\left(\dfrac{36\ MHz}{TPE}\right)} \right\rceil = 50 \quad (7.2)$$

If the ownership costs are as follows:

$$\text{Cost of government satellite} = \$400\text{M}$$

$$\text{Cost of launch vehicle} = \$150\text{M}$$

Then the cost per TPE per year required to achieve a cost neutral position is found using Equation (7.3):

$$Cost(\$M \; per \; TPE \; per \; year) = \frac{400 + 150}{(50)(15)} = \$0.73M \; per \; TPE \; per \; year \quad (7.3)$$

Thus, the price per TPE per year must be less than $730K for cost equivalency to the owned asset. If instead the commercial transponder lease price is actually $1.5M/TPE/year, the additional cost of diversifying using commercial services in this example is calculated using Equation (7.4):

$$Added \; Resilience \; Cost = \left(1800 \; MHz / \left(36 \; MHz / TPE\right)\right)$$
$$\left(\$1.5M - \$0.73M\right) = \$38.3M / yr \quad (7.4)$$

Since this approach provided a 15 percent increase in resilience, the cost per percentage of resilience gained is $38.3M/15 = $2.6M/year. This cost must be balanced against other mitigation approaches.

If the government satellites can be made more robust to the Ka-band jammers, the value of diversification is reduced. Figure 7.9 shows how the system-level resilience benefit decreases for this example as the individual government satellites are made more resilient to the jamming threat. If the satellite resilience can be improved to from 0.4 to 0.8, the added benefit of this diversification approach decreases to only about a 5 percent improvement for the same cost of diversification. If the cost of developing the technology

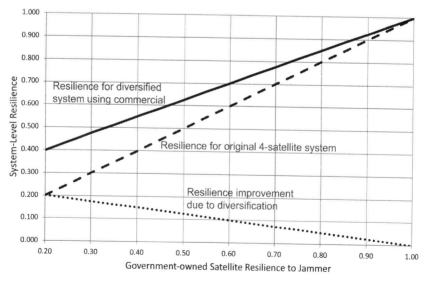

**FIGURE 7.9**
Resilience benefit due to diversification versus elemental resilience.

to accomplish this, plus the cost of implementing it on the four government satellites, and any additional costs to the ground and user segments is significantly less than the total cost of diversification (Equation (7.5)):

$$\text{Cost of diversification} = (\$38.3\text{M}/\text{yr})(15 \text{ yrs}) = \$575\text{M}. \qquad (7.5)$$

then this approach might be more cost effective, assuming other factors such as performance and risk are also acceptable, even though the resilience is still 10 percent less.

But if an adversary responds by acquiring C-band jammers to interfere with the commercial services as well, then the threat has changed once again and the system-level resilience is reduced. An advantage of using commercial services in this example is that they can be more agilely purchased "by the yard" in small capacity increments across a large number of satellites and quickly dispensed with when no longer needed. As such, there is less risk in making large investments to infrastructure that could be countered over the system's life span. On the other hand, for large, enduring capacity demands over a long period of time, cost of ownership might be the lower cost alternative.

## 7.4 Increasing Resilience Through Responsive Recovery and Diversification

The use of existing commercial services can also provide another avenue for obtaining resilience, this time through responsive recovery. If the services of a government satellite in the previous example were to be disrupted by a jammer and some portion of the users are equipped with terminals with dual-band capability, meaning that the terminals can be operated at both military and commercial frequencies, then rapid recovery of the service might be possible by activating commercial services and moving some amount of lost user traffic to those commercial satellite transponders. Though similar to the previous example, this entails making a responsive change to the operational system configuration when confronted by a threat. Figure 7.10 illustrates a reassignment of 50 percent of the lost Ka-band traffic due to interference to commercial assets, resulting in a loss of only $(0.75)(0.6)$ $(0.5) = 0.225$, or 22.5 percent of the total capacity, for a resilience of $R = 0.775$. The resilience value may be somewhat less when accounting for recovery time, as discussed in Chapter 5.

The ability to move from a government system to a commercial service is dependent on a number of factors in addition to the availability of a user terminal that can access both. The performance of the two satellites must also be comparable or else there could be a reduction in data rates, bandwidth,

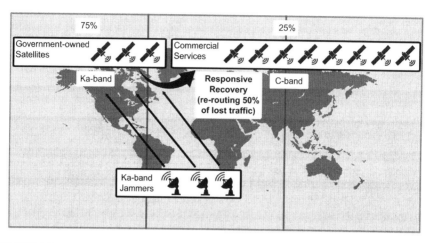

**FIGURE 7.10**
Use of responsive recovery to improve resilience to jamming.

and/or capacity. The commercial service(s) must be available "on call," which means that the satellite resource might need to be pre-coordinated. (Note that following responsive recovery the percentage of system capacity carried by the commercial services has risen from 25 to 47.5 percent.) As a result, the amount of traffic that can be re-routed varies depending on all of these factors. This arrangement brings with it a cost either to reserve the required commercial bandwidth for when it is needed, or to potentially pay a premium to preempt other users; in either case there will be some additional cost.

Though this example features a communications system, other space systems would have the same trades. A high-performance imaging system could in some way be disabled, but a secondary commercial system could be accessed, although with some sacrifice of image quality. The commercial imagery would likewise be provided as a commercial service and could be used in conjunction with the government-owned system, as with the communications example.

Again, assuming an adversary is intent upon jamming military frequency bands, this responsive reconfiguration to access commercial services could prove an adequate fallback position and the services could be dispensed with at a later time if the jamming threat is neutralized or is otherwise not persistent. This is an example of a system being both reconfigurable and agile in achieving increased resilience. This is not a trivial approach, though, as the proper user terminals must be established with fielded, predefined CONOPS, and commercial operator service agreements must be pre-arranged. The degree to which this is possible determines how quickly user traffic can be reestablished using this approach.

A drawback to this type of architecture was already illustrated in Figure 6.9, which shows the effect on system-level resilience of deviating from a fully uniform architecture. Across the threat range, non-uniform architectures feature regions of decreased resilience across the threat range. In this example, the jamming threat is pervasive, with a jammer for each government satellite. For other threats, the larger high-value assets attract the most attention and successful attacks will disproportionately reduce the system capability compared to the uniformly distributed architecture. Nevertheless, this diversification approach has clear resilience benefits if it is practical to implement.

Using commercial services in this example might also provide some additional resilience against non-electronic threats due to the fact that they are commercial assets and less likely to be targeted due to their commercial value and the fact that many different customers benefit from the satellite (perhaps including the adversary). All of these factors must be considered in choosing any commercial assets, including the trustworthiness of the owners, and users of the individual satellites. In this example, eight different commercial satellites are used to provide the aggregate service, and this also provides some measure of resilience against multiple types of threats by further distributing the capability, as discussed in detail in Chapter 6. In the example, each commercial satellite carries only about 3 percent of the total system capacity under benign conditions.

## 7.5 Designing for Resilience to Multiple Threats

The next scenario examines system architectures that provide varying levels of resilience against two distinct threats to the space segment: a physical threat and an electronic threat. In this example, both threats are directed at the same elements of the system and so the combined effects must be examined so as to correctly calculate the total system-level resilience.

Consider the following scenario: A military space system provides secure global communications using four identical satellites each carrying 25 percent of the total global capacity (Figure 7.11). The minimum system resilience is $R_{RQ} = 0.4$. The system must be resilient to two distinct threats:

1. A kinetic threat targeting all four satellites. The $P_k$ for each of the ASATs is 0.75, so each attack has a 75 percent probability of success given no threat mitigations.

2. An electronic threat that is directed at all four satellites. Each satellite experiences a capacity loss of 50 percent when subjected to this jammer, for an elemental resilience of 0.5.

Evaluating the resilience of this system to the kinetic threat, with no avoidance, recovery, or reconstitution, the system-level robustness is the only contributor and is found by calculating the expected value of residual capacity when all four satellites are attacked.

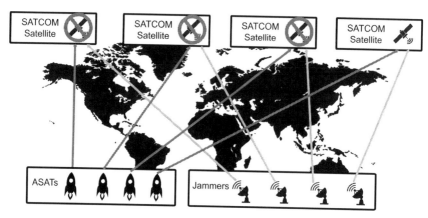

**FIGURE 7.11**
Baseline four-satellite GEO system with multiple threats.

Using a binomial probability calculation to find the resilience to the kinetic threat based on the expected number of surviving satellites:

Resilience to the kinetic threat $R_k = R_{RO} = (\text{Number of surviving satellites})/4$

$$\text{Expected number of surviving satellites} =$$
$$\begin{pmatrix} 4*P(\text{loss of } 0) + 3*(P(\text{loss of } 1)*4) \\ + 2*(P(\text{loss of } 2)*6) + P(\text{loss of } 3)*4 \end{pmatrix} = 1$$

One satellite is expected to survive (loss of three), so the kinetic resilience $R_k = (1/4) = 0.25$. This does not meet the minimum required value of 0.4, and so this architecture is insufficient to meet its required resilience against the kinetic threat.

Resilience to the electronic threat, $R_e = 0.5$, as each satellite's capacity is reduced by that factor, and is compliant. But when the two threats are considered together, then the overall resilience is the product of the two separate resilience values, as the single remaining satellite's capacity is further degraded by the second, electronic threat:

$$R = (R_k)(R_e) = (0.25)(0.5) = \mathbf{0.125}, \text{ which is also non-compliant.}$$

This initial system design does not meet the resilience requirements so additional mitigations are required. As the kinetic resilience represents the largest loss factor, the simplest response is to distribute the capability among a larger number of satellites. From Chapter 6, Equation (6.2), the number of *additional* uniform satellites required to meet the kinetic resilience require-ment can be found from:

$$M \geq \frac{4}{1-0.4} - 4 = 2.67$$

Rounding up to M = 3, the total number of satellites becomes N + M = 4 + 3 = 7, and losing 3 (as before) results in $R_k = (4/7) = 0.57$. Using the same electronic resilience value of 0.5, the new system-level resilience becomes R = (0.57)(0.5) = 0.286, which is better but still not compliant.

If instead additional *satellites of the same size* are added, employing prolif-eration instead, the number required can be found from Equation (7.6):

$$M \geq N\left(R_{RQ} - 1\right) + TP_k \tag{7.6}$$

so M ≥ (4)(0.4 – 1) + (4)(0.75) = 0.6 (rounded up to 1 additional satellite)

In this case, only a single additional satellite is required to be added to meet the system-level capacity required, for a total of five satellites, because more capacity has been added to the system. The kinetic resilience $R_k = (2/5)$ = 0.4, however, since 25 percent margin has been added to the system, the required resilience has been reduced to $R_{RQ} = 0.4/1.25 = 0.32$ to meet the orig-inal requirement (capacity of four satellites). When the electronic resilience is included, R = (0.4)(0.5) = 0.2, so this approach is also non-compliant despite improving the kinetic resilience. Thus additional steps must be taken for either approach to meet the system-level requirement.

Another solution is to add threat mitigation features to increase the ele-mental resilience of each of the satellites to enable them to better avoid the kinetic threat. If these satellite protection features can increase the avoidance from the original $R_{AV} = 1 - P_k = 0.25$ to the much higher value of 0.95, then the robustness of the four-satellite system increases to $R_k = 0.95$, far surpass-ing the required value of 0.4. Even if the resilience to the electronic threat remains at 0.5, the overall system resilience is R = (0.95)(0.5) = 0.475, which meets the requirement of 0.4.

A third approach is to adopt commercial diversification as described in the previous example. For this alternate architecture (Figure 7.12), 65 per-cent of the total capacity is provided equally by the four government-owned satellites while 35 percent is provided equally by six commercial satellites. The kinetic threat still targets four satellites (T = 4). If an adversary now chooses to exercise the kinetic threat, it is to its advantage to do so against the satellites carrying the highest capacity, which are the government sat-ellites that each carry almost triple the capacity of each of the commercial satellites [(0.65/4) versus (0.35/6)]. For this scenario the assumption is that the electronic threat can be deployed against *all* of the satellites in use, both government and commercial (using the appropriate transmit frequencies).

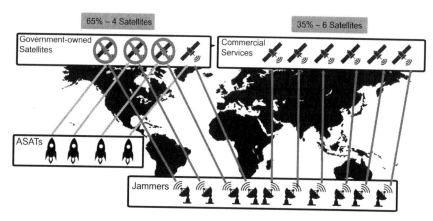

**FIGURE 7.12**
Multiple threat scenario for hybrid architecture.

Now, as before, without mitigation three of the four government satellites targeted by the kinetic weapons are projected to be disabled. But none of the commercial satellites are targeted, so they experience only the electronic threat. The kinetic resilience of this architecture is thus:

$$R_k = (0.65)(0.25) + (0.35)(1) = 0.51$$

meeting the resilience requirement of 0.4.

But in this scenario the commercial satellites are assumed to be vulnerable to the electronic threat, losing all capacity when targeted. The resilience to the electronic threat alone is:

$$R_e = (0.65)(0.5) + (0.35)(0) = 0.325$$

Against the combined threat for this architecture, the resilience is:

$$R = (0.65)\left[\left(R_{k,Gov't}\right)\left(R_{e,Gov't}\right)\right] + (0.35)\left[\left(R_{k,Comm'l}\right)\left(R_{e,Comm'l}\right)\right]$$
$$= (0.65)(0.25)(0.5) + (0.35)(1)(0) = \textbf{0.081}$$

While this change in architecture improved the resilience solely to a kinetic threat, it reduced the resilience to both an electronic threat and, by extension, to the combined threat.

In this case, a combination of both distribution and protection resilience features might be required to achieve the desired resilience. If the anti-jam protection for the government satellites could be raised from 0.5 to 0.9 and alternate commercial systems are selected to maximize robustness against the jammers, perhaps using smaller antenna beams to also raise the robustness to 0.9, then the resulting overall resilience becomes:

$$R = (0.65)(0.25)(0.9) + (0.35)(1)(0.9) = \mathbf{0.46}$$

This approach, if feasible, would meet the original system resilience requirement of 0.4, albeit with a modified threat scenario.

## 7.6 Designing for Resilience and Cost in a Multi-Threat Environment

So far, the examples provided have been greatly simplified and focused on the resilience trade to illustrate particular aspects of the design trade space for space system architectures. The third trade parameter, cost, must also be included in a real-world design activity. In the following example, the designer is provided a library of building blocks that are available for use in constructing the space layer. Each of these blocks provides some amount of capability (capacity), elemental resilience to two threats, and an associated cost. The goal is to use this library to construct a fully compliant space layer that meets the cost, performance, and resilience requirements.

The requirements for a SATCOM system are as follows:

- Minimum capacity: 20 Gbps (to a defined coverage area)
- Minimum system resilience: $R_{RQ} = 0.75$
- Maximum space layer cost: $1B (15-year lifetime)
- Threats:
  - Kinetic threat: Maximum of two ASATs
    - ASAT effectiveness: 0.80 (in the absence of countermeasures)
  - Electronic threat: Maximum of 10 jammers (at the appropriate frequency)

Figure 7.13 shows the available space layer options, both satellites and hosted payloads, government-owned and commercial services. Each of these building blocks provides a specific amount of capacity and varying degrees of resilience to the electronic and kinetic threats. These resilience properties are expressed in terms of robustness and recovery values to the electronic threat. The effectiveness of the kinetic threat is also shown: no on-board countermeasures are assumed and hosted payloads are not considered targets. In this example, there is no option for reconstitution.

| Satellite Type | Capacity (Gbps) | Gov't/ Commercial | Cost ($M) | Electronic Threat | | Kinetic Threat |
| --- | --- | --- | --- | --- | --- | --- |
| | | | | Robustness | Recovery | Effectiveness |
| Small Protected Free Flyer | 2 | G | 250 | 0.80 | 0.75 | 0.8 |
| Small Hosted Payload | 2 | G | 100 | 0.40 | 0.00 | 0 |
| Large Transponded Wideband | 5 | G | 500 | 0.40 | 0.00 | 0.8 |
| Commercial HTS | C | C | 0.6/TPE/yr | 0.25 | 0.00 | 0.8 |
| Commercial FSS | C | C | 1.5/TPE/yr | 0.00 | 0.00 | 0.8 |

**FIGURE 7.13**
Space layer satellite design options.

Three different system designs for the space layer are created and evaluated to determine which best meets the performance, cost, and resilience requirements using the available options:

1. Design #1 (Figure 7.14) meets the capacity requirement through the use of three large wideband government satellites, each providing 5 Gbps of capacity, and three small protected wideband government satellites, each providing 2 Gbps, for a total of 21 Gbps. *Since excess margin is included, the adjusted required resilience specification is:*

   $R_{RQ} = (0.75)(20/21) = 0.714$, *as* $(0.75)(20) = 15$ *Gbps is still the required system capacity.* The smaller satellites have features that provide more anti-jam robustness against the electronic threat.

2. Design #2 (Figure 7.15) employs 10 small protected wideband government satellites, each providing 2 Gbps, for a total of 20 Gbps.

3. Design #3 (Figure 7.16) is a system with two large wideband government satellites (5 Gbps), two small protected wideband government satellites (2 Gbps each), and 6 Gbps of services from one commercial high throughput satellite (HTS), for a total capacity of 20 Gbps.

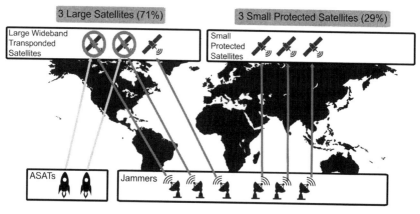

**FIGURE 7.14**
Design #1: Three large and three small satellites.

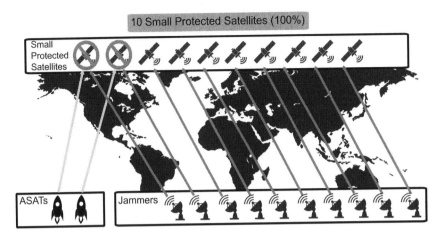

**FIGURE 7.15**
Design #2: Ten small satellites.

In this example the performance, which is the system capacity, has already been calculated as part of the design process in establishing the three system options above. The next step is to calculate the resilience for each design option against each threat.

## 7.6.1 Design #1 Resilience Calculation

*Kinetic resilience:* The two ASATs are assumed to be targeting two of the three large satellites, and the ASAT effectiveness is 0.80. It is convenient to first calculate the resilience of the large satellites as a group, as they are the only ones being targeted and therefore the effective resilience of the smaller satellites is 1.

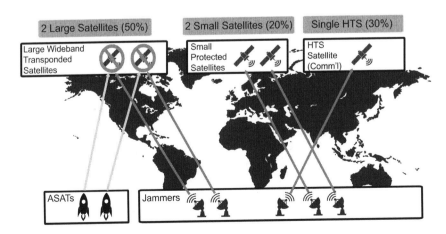

**FIGURE 7.16**
Design #3: Hybrid government/commercial architecture.

As shown previously, targeting N satellites each with an avoidance of $R_{AV}$ yields a resilience of $R_{AV}$. In this case $R_{AV} = 0.2$, so the resilience of the two targeted satellites to the threat is 0.2 and the kinetic resilience of the three taken together can be found by:

$$R_k \left(\text{Large}\right) = (0.2)(2/3) + (1/3) = 0.467$$

When adding in the contribution of the three additional smaller satellites which are unaffected, and weighting according to the capacity of each group, the resilience to the kinetic threat is:

$$R_k = (0.467)(3)(5/21) + (1)(3)(2/21) = 0.619$$

This resilience to the kinetic threat alone is less than the required 0.714 and is non-compliant.

*Electronic threat:* The electronic threat cannot be avoided, so the robustness and recovery values (from Figure 7.13) for each targeted asset must be used to calculate the resilience to this threat. In this case, the small protected satellite provides less capacity but maintains it more effectively when jammed because each of the small protected satellites has a robustness of 0.8 and recovery of 0.75. The resilience can be found from applying the capstone equation of Figure 5.2 for robustness and recovery:

$$R_e \left(\text{Large}\right) = (0.40) + (1 - 0.4)(0) = 0.40$$

$$R_e \left(\text{Small protected}\right) = (0.80) + (1 - 0.8)(0.75) = 0.95$$

Using the same approach as for the kinetic threat and weighting each contributor according to its fractional capacity contribution, the resilience to electronic threat is:

$$R_e = (0.40)(3)(5/21) + (0.95)(3)(2/21) = 0.557$$

In the case of the combined threat scenario, with both kinetic and electronic threats present, the resilience is the result of a combination of effects. Examining the case of the kinetic threat, there is a high probability (64 percent) that the two satellites targeted by the ASATs are disabled, a 32 percent probability that one is disabled, and only a 4 percent probability that neither is disabled. Taking a conservative approach, we assume the most likely scenario that two of the three large satellites are disabled. Now the effect of electronic jamming can be applied to the resulting satellite configuration, with the new combined space-layer resilience calculated by:

$$R = (1)(5/21)(0.4) + (3)(2/21)(0.95) = \mathbf{0.367}$$

Once again, this value is much lower than the adjusted required resilience of 0.714.

## 7.6.2 Design #2 Resilience Calculation

*Kinetic resilience:* There is a single group of 10 small protected satellites. Two are targeted with threat effectiveness of 0.8. The resilience is again equal to $R_{AV} = 0.2$ for these satellites. The space-layer resilience is found by:

$$R_k = (0.2)(2)(2/20) + (1)(8)(2/20) = 0.84$$

*Electronic resilience:*

$$R_e(\text{Small protected}) = (1)(0.80) + (1)(1 - 0.8)(0.75) = 0.95$$

The combined resilience to kinetic and electronic threats (using the same worst case as Design #1, with a loss of two satellites):

$$R = (0.95)(8)(2/20) = \mathbf{0.76}$$

This design solution meets the resilience requirement of $R_{RQ} = 0.75$.

## 7.6.3 Design #3 Resilience Calculation

*Kinetic resilience:* This is similar to Design #1, but only two large wideband satellites are in this design and 30 percent of the capacity is delivered using a commercially leased service over an HTS satellite. Again the assumption is that the two large wideband satellites are targeted by the kinetic threat:

$$R_k(\text{Large}) = R_{AV} = 0.2$$

$$R_k(\text{Small protected}) = 1$$

$$R_k(\text{HTS}) = 1$$

$$R_k = (0.2)(2)(5/20) + (1)(2)(2/20) + (1)(6/20) = 0.60$$

*Electronic resilience:*

$$R_e(\text{Large}) = 0.4$$

$$R_e(\text{Small protected}) = (1)(0.80) + (1)(1 - 0.8)(0.75) = 0.95$$

$$R_e(\text{HTS}) = 0.25$$

$$R_e = (0.4)(2)(5/20) + (0.95)(2)(2/20) + (0.25)(1)(6/20) = 0.465$$

*Combined resilience:*

$$R = (0.4)(0) + (0.95)(4/20) + (0.25)(6/20) = \mathbf{0.265}$$

Again, this option falls significantly short of the required resilience. From a resilience viewpoint, Design #2 is the only one that meets the requirement. The next step is to calculate the cost of each design using values from Figure 7.13:

Design #1 space-layer cost = (3)(500) + (3)(250) = $2.25B, *above the $1B budget*

Design #2 space-layer cost = (10)(250) = $2.5B, *also above the target cost*

Design #3 space-layer cost = (2)(500) + (2)(250) + 1500 = $3B, *the highest cost option*

Figure 7.17 summarizes the performance of each of the three designs versus the requirements and illustrates how three different alternatives, each with similar performance (capacity) and cost, could have wildly different resilience. The first table shows the absolute values as presented above. The second table contains these values normalized to a scale of 0 to 10 for the purpose of better graphically displaying the comparison among the three design options. These values are plotted on a radar or spider chart, as shown in Figure 7.18. This graphical method gives a more visual view of the key results in a single plot. The values have been normalized such that 10 is always best, and 0 worst. The goal is to obtain a plot with the greatest area among all four variables. In this case, all three options provide a high value for capacity, as that was a key design-to requirement. None provide a good cost value relative to the budget. Design #2, as noted previously, provides improved resilience to both threats. Similarly the combined system resilience could have been included.

Clearly none of these three designs meets the cost target for a 20 Gbps capacity system, with the lowest cost being more than double the target. Design #2 provides the greatest resilience per unit cost. Other costs, including launch,

Absolute Values

| Option | Capacity (Gbps) | Resilience to Kinetic | Resilience to Electronic | Cost ($B) |
|---|---|---|---|---|
| Design #1 | 21 | 0.62 | 0.56 | 2.25 |
| Design #2 | 20 | 0.84 | 0.95 | 2.50 |
| Design #3 | 20 | 0.60 | 0.47 | 3.00 |

Normalized to Scale of Max 10

| Option | Capacity | Resilience to Kinetic | Resilience to Electronic | Cost |
|---|---|---|---|---|
| Design #1 | 10 | 6 | 5 | 4.00 |
| Design #2 | 9 | 8 | 9 | 3.00 |
| Design #3 | 9 | 6 | 4 | 2.00 |

**FIGURE 7.17**
Summary of design options.

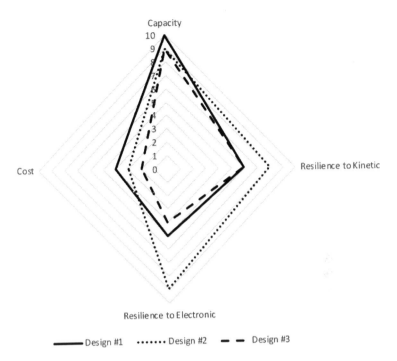

**FIGURE 7.18**
Radar graph comparing three options.

ground, and user terminals, will also influence the cost trade. For example, Design #2 could require more launches, and the aggregate launch cost could push the total cost above Design #1's cost. But if multiple smaller satellites can be launched on a single launch vehicle, the launch costs will be more comparable.

In most cases, the entire system cost must be considered in the trade over the period of performance or on an average annual basis. Sometimes this is referred to as the life-cycle cost (LCC) or the enterprise cost. As complexity of space systems has increased, more of the cost has become located in the ground segment, including user terminals for communications systems. Thus, simply examining costs for the space segment is insufficient in obtaining the true system trade cost. Additionally, the development costs must be included and are often amortized across the cost of a block of satellites or ground stations.

Another source of cost comes from any unique threat mitigation features included in the system. For example, in the above example, Design #3 relies on a smaller number of large satellites. As has been discussed before, the opportunity for on-board countermeasures for one or more satellites can provide increased resilience. If a countermeasure can increase the resilience for the two large wideband satellites to 0.50, the kinetic resilience would increase to 0.75. Similarly, additional anti-jam measures might also be used to improve the resilience to the electronic threat as well.

This example and the previous one highlight how a single threat mitigation technique is likely insufficient against multiple threats, in contrast to making architectural changes such as distribution or diversification which can increase the robustness to multiple threats. The challenge is how to best select the combination of mitigation features to minimize the increased cost to the system. In this example, Design #1 was the most affordable option but fell significantly short of the resilience requirement. Redistributing some of the capacity from the larger satellites to the smaller, protected ones could improve the resilience to the electronic threat with minimal cost penalty, but with a slight loss of capacity. In this revised case, there are now two large satellites and five smaller protected satellites for a total capacity of 20 Gbps (which still meets the requirement).

$$Kinetic\ resilience: R_k = (0.2)(2)(5/20) + (1)(5)(2/20) = 0.60$$

$$Electronic\ resilience: R_e = (0.4)(2)(5/20) + (0.95)(5)(2/20) = 0.675$$

$$Overall\ resilience\left(still\ losing\ the\ two\ large\ satellites\right):$$
$$R = (0.95)(5)(2/20) = \mathbf{0.475}$$

In this case, by simply moving capacity from one type of asset to another, the kinetic resilience is maintained, but the electronic and overall resilience is significantly increased, with no net satellite cost increase (one $500M satellite was replaced by two $250M satellites). The net result is that 1 Gbps of capacity was traded for an increased resilience of about 11 percent. Though this exercise did not produce a compliant design, it shows how performance, resilience, and cost can be traded to achieve an optimal solution.

## 7.7 Multiple Threats and Multiple Mitigations Example

Real-world resilient space system designs are much more complex than the examples provided so far. The following example includes even more detail to highlight the multiple levels of trades that must be considered when designing a system from a clean sheet of paper. Costs are now allocated down to the feature level and also associated directly with relative satellite size. Launch costs are included, as well as opportunities for multi-satellite launch based on size. Certain on-board mitigation features are not available for certain sized satellites. Finally, the cost of the user terminals is also included in the total system life-cycle cost and must be compatible with the satellites that the user(s) will access.

Once again, multiple system designs are created from these building blocks in an attempt to meet performance, resilience, and cost requirements.

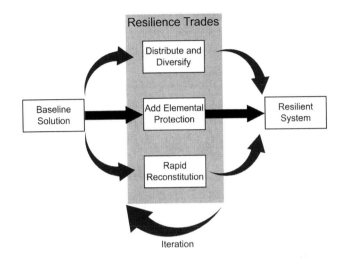

**FIGURE 7.19**
Resilience trade space includes multiple mitigation approaches.

These alternative designs are then compared to determine which best meets the requirements. The difference is the level at which the optimization can occur and the number of available sub-options in creating the system design.

In addition, a wider range of threat mitigation features are examined. Figure 7.19 shows the trade space, which includes previously discussed distribution and diversification mitigations, elemental protection (on-board countermeasures), and also rapid reconstitution of entire satellites. As before, multiple system design alternatives are created and traded in an attempt to optimize the system and meet the required resilience. As these alternatives are considered and evaluated, multiple iterations may be required to optimize the system design.

The system level requirements are as follows:

1. Total system communications capacity of at least 30 Gbps, covering the Earth with equal amounts of capacity per region (10 Gbps per 120° of longitude).
2. Fifty percent of the capacity of each satellite must be routed via a ground gateway site.
3. Minimum system level resilience of $R_{RQ} = 0.75$ in the presence of all threats.
4. Threats:
   a. Up to 5 RF jammers in the military Ka frequency band
   b. Kinetic ASAT threat of up to 3 ASATs with effectiveness of $P_k = 0.9$
   c. Physical threat to a single ground site with effectiveness of 1

| | | | Electronic Threat | Kinetic Threat | | | |
|---|---|---|---|---|---|---|---|
| Satellite Type | Capacity (Gbps) | Frequency Band | Robustness | Effectiveness | Satellite Cost ($M) | Launch Cost ($M) | Ground Cost (M) |
| **Space Segment** | | | | | | | |
| Hosted Payload | 0.5 | Mil Ka | 0.10 | 0.9 | 75 | 0 | |
| Small Satellite | 1 | Mil Ka | 0.25 | 0.9 | 100 | 45* | |
| Medium Satellite | 2.5 | Mil Ka | 0.30 | 0.9 | 200 | 90 | |
| Large Satellite | 5 | Mil Ka | 0.50 | 0.9 | 300 | 90 | |
| High Capacity Satellite | 10 | Mil Ka | 0.50 | 0.9 | 450 | 150 | |
| **Ground Segment** | | | | | | | |
| Gateway, 2 antennas | 10 | Mil Ka | | | | | 50 |
| Gateway, 4 antennas | 10 | Mil Ka | | | | | 65 |

\* Dual satellite launch

**FIGURE 7.20**
Available classes of satellites and ground gateways.

5. The available satellite options for the space and ground segments and their characteristics are shown in the table in Figure 7.20.

6. The available threat mitigation options are shown in Figure 7.21, including costs for each.

Once again, a baseline system is created using as simple an architecture as possible to meet the key performance requirements. This architecture is shown in Figure 7.22 and is based on three large high-capacity GEO satellites with characteristics as shown in Figure 7.20, each capable of delivering 10 Gbps of capacity to its coverage area. A minimum of three ground gateway sites are required, and for added resilience and reliability each site can support two satellites simultaneously (10 Gbps per link). Each satellite can see two ground sites, though only a single communications link is active at one time.

In the absence of any threat mitigation features for either space or ground, the overall system level resilience is low due to its centralized nature and lack of significant margin and/or redundancy. Examining each threat's independent impact in turn:

*Kinetic:* One ASAT against each of the three satellites yields $R_k = 0.10$

*Electronic:* A single jammer against each satellite results in $R_e = 0.5$

*Physical:* One ground gateway is disabled, no loss of capacity, $R_p = 1$

| Threat Mitigation | Threat Type | Avoidance | Robustness | Recovery | Reconstitution | Development Cost ($M) | Recurring Cost ($M) |
|---|---|---|---|---|---|---|---|
| Satellite Maneuverability | Kinetic - Space | 0.95 | | | | 100 | 35 |
| Enhanced Anti-Jam Payload | Electronic - Space | | 0.8 | 0.75 | | 80 | 60 |
| Rapid Launch of Satellite | Multiple | | | | 0.75 | 0 | 400 |
| Added Redundancy | Electronic - Space | | | 0.25 | | 5 | 5 |

**FIGURE 7.21**
Available threat mitigation options.

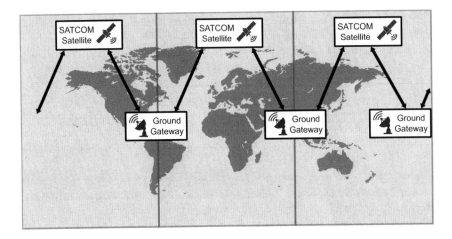

**FIGURE 7.22**
Baseline three-satellite space system.

There is a high likelihood that all three satellites are disabled due to the kinetic threat, so it is reasonable to simplify and assume none survive without mitigation. But from a mathematical perspective, a worst-case resilience value due to the combined effects of all threats can still be calculated to be:

$$R = (R_k)(R_e)(R_p) = (0.1)(0.5)(1) = \textbf{0.05}$$

Therefore, mitigations must be put in place to meet the resilience requirement. Three different approaches are examined: elemental protection, distribution and diversification, and rapid reconstitution.

*Option 1: Protection of Satellites* The satellite maneuverability countermeasure for the kinetic threat (Figure 7.21), provides an avoidance $R_{AV} = 0.95$. Using Equation (5.9), with $P_E = 0.9$, $P_D = 1$, and $P_C = 0.95$, $R_{AV} = 0.96$ for a single satellite. When three are attacked simultaneously by the same threat, the system resilience to the kinetic threat is also $R_k = 0.96$. This result indicates high confidence (probability of 88.5 percent) in all three satellites surviving.

Applying the electronic threat, the combined resilience is reduced to:

$$R = (0.96)(0.5) = \textbf{0.48}$$

The list of available mitigations also includes additional anti-jam resilience through both added robustness ($R_{RO} = 0.8$) and recovery ($R_{RV} = 0.75$). The adjusted resilience to the electronic threat with this mitigation feature thus becomes $R_e = 0.8 + (1 - 0.8)(0.75) = 0.95$. Now the combined resilience to both kinetic and electronic threats with both mitigations increases to:

$$R = (0.96)(0.95) = \textbf{0.91}$$

The loss of a single ground gateway has very little impact in this scenario so long as each satellite can responsively redirect its traffic to the second site in its field of view to recover the lost link. For this example, this is performed as a pre-programmed procedure resulting in very limited interruption of service. As a result, the resilience to all three threats using the two mitigations to add elemental protection to meet resilience requirements is 0.91, which provides sufficient margin above the requirement of 0.75.

***Option 2: Distribution and Diversification*** The second approach is to distribute and/or diversify the space layer as shown in previous examples. In this example, there are more choices and a greater ability to optimize the system versus cost and resilience, particularly in terms of sizing the satellites to meet a desired capacity. To start, assume no additional elemental mitigations are being used.

The initial number of elements (N) is 3, so the minimum number of additional elements (M) required to meet the kinetic resilience for T = 3 (loss of three satellites) can be found from Equation (6.2):

$$M \geq \frac{3}{1-0.75} - 3 = 9$$

M = 9, so N + M = 3 + 9 = 12 satellites total. This corresponds to the "Medium satellite" shown in Figure 7.20, with a capacity of 30/12 = 2.5 Gbps. This architecture is shown in Figure 7.23 and includes three additional ground gateways to support the increased number of satellites. To minimize new sites (to six), the number of antennas at each site has been increased from two to four, with associated cost per Figure 7.20. The resulting performance to the kinetic threat now can be validated by calculating the resilience of a 12-satellite architecture to the threat T = 3. In the worst case, for a threat

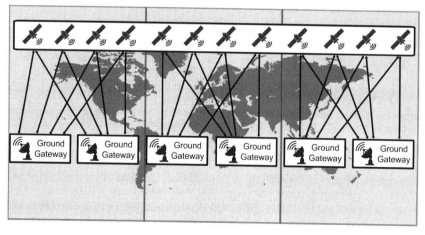

**FIGURE 7.23**
Option 2 distributed 12-satellite 6-gateway architecture.

effectiveness of 1, 9 of the 12 satellites would remain, for a resilience of (9/12) = 0.75. In this case, the threat effectiveness is 0.9, so the calculated kinetic resilience is slightly higher:

$$R_k = ((0.1)(3) + 9)/12 = 0.775$$

And the resilience solely to the electronic threat is simply:

$$R_e = 0.5$$

Again assuming a worst-case loss of the three targeted satellites, five of the nine remaining satellites following the kinetic attack are targeted by the five RF jammers, resulting in the following resilience to jamming relative to the original 12 satellite constellation. The combined resilience is:

$$R = ((0.5)(5) + 4)/12 = \mathbf{0.54}$$

As in the protection approach, the electronic threat mitigation could also be used to raise the resilience from 0.5 to 0.95, resulting in a combined resilience of:

$$R = ((0.95)(5) + 4)/12 = \mathbf{0.73}$$

which almost meets the 0.75 requirement; adding one more satellite raises this value to 0.75 (versus an adjusted required resilience for 13 satellites of 0.75/(13/12) = 0.69). A drawback is that now this mitigation must be installed on 12 (or 13) satellites rather than only three, as there is no way to know which of the satellites will be targeted by a jammer, particularly for a uniformly distributed system. This results in an increased mitigation cost for the system.

Using diversification instead can be a more cost-effective approach to avoid at least some of the jamming impact. If half of the 12 satellites are providing capacity through commercial services and are not subject to kinetic targeting nor vulnerable to the military frequency jammers, then the resilience increases:

$$R = ((0.5)(3) + 6)/12 = \mathbf{0.625}$$

While this does not meet the resilience requirement, this again illustrates how reallocation of military and commercial capacity among the two types of satellites might result in higher resilience against the combined threat.

*Option 3: Rapid Reconstitution* A third mitigation approach is the use of rapid reconstitution to quickly replace capacity through the responsive launch of new satellites to replace those disabled. In this case, the entire lost capability is replaced, with the resilience being more a valuation of the time to reconstitute. This valuation must be consistent with other resilience values in order to assure an accurate trade. Figure 7.24 shows a notional value table for the reconstitution metric based on this time. Using this table, if a replacement satellite can be operational within one week of callup, $R_{RC} = 0.75$.

| Time to Reconstitute | $R_{RC}$ Value |
|:---:|:---:|
| 1 day | 1 |
| 3 days | 0.9 |
| 1 week | 0.75 |
| 2 weeks | 0.5 |
| 3 weeks | 0.3 |
| 1 month | 0.4 |
| 2 months | 0.2 |
| 3+ months | 0 |

**FIGURE 7.24**
Reconstitution value versus time.

Reconstitution changes the system-level resilience value depending upon the architecture of the space layer and how many elements (satellites) are replaced. For the baseline architecture with three high-capacity satellites, if all three are lost ($R_k$ was found to be 0.1) and all three can be reconstituted within one week, with the electronic resilience still 0.95, the baseline resilience is:

$$R = (0.1)(0.95) = \mathbf{0.095}$$

With reconstitution, again using the equation in Figure 5.2, the system resilience becomes:

$$R = 0.095 + (1 - 0.095)(0.75) = \mathbf{0.774}$$

If the second architecture with 12 satellites is the starting point, and three are lost, then the new resilience for reconstitution of all three satellites is:

$$R = 0.73 + (1 - 0.73)(0.75) = \mathbf{0.93}$$

Admittedly, accomplishing rapid reconstitution on short timelines is a difficult and potentially expensive proposition. Additional satellites and launch vehicles must be built and readied for responsive callup at a launch site, affecting existing schedules. This requires extraordinary effort to expedite this kind of activity but could conceivably be accomplished with sufficient planning and investment. Physics could become a limiting factor in getting a satellite into its orbital position quickly as well, as the time to place a satellite into a GEO orbit can take weeks or months depending upon the launch vehicle capability and the satellite's mass. With a wider range of launch vehicles available at reduced prices the launch paradigm could change significantly, opening up such alternatives in the future. And for emerging highly distributed systems of many small satellites (smallsats), this approach could become the norm as the launch pace may already support it.

In practice, the value of reconstitution is likely driven by comparisons of its value to other available mitigations, such as recovery options. If avoidance, robustness, and recovery still result in considerable capability shortfall, then the only way to achieve an acceptable level is through reconstitution, so that influences its value significantly. The relative costs also enter into the trade. One conclusion is that systems with very high resilience requirements favor reconstitution if other methods fall short. In this example, on-board mitigations have been shown to provide ample mitigation to achieve the 0.75 system resilience, meaning that unless reconstitution is much more affordable, it will likely not be competitive with these other approaches.

In the baseline architecture, on-board kinetic and electronic mitigations resulted in $R=0.91$, which is more than adequate. Had all mitigations resulted in a large shortfall, reconstitution might be more attractive if affordable.

Reconstitution is thus another option for obtaining increased resilience when the others are either insufficiently effective, or unaffordable. Yet another reason is when existing on-orbit mitigations cease to be effective and the satellite must be refreshed to maintain the desired resilience. This might be due to threat escalation or loss of functionality. At that point the resilience could decrease to the point that a known threat could cause unacceptable service loss and must be replaced.

*Cost comparison* The final step is to compare the costs of each of the evaluated approaches. These non-recurring and recurring costs are taken from the notional costs shown previously in Figures 7.20 and 7.21.

**Option 1:** The cost of the three high-capacity satellites and associated launch vehicles is:

$$Cost = (3)(\$450M + \$150M) = \$1.8B$$

The added cost of mitigating the kinetic threat is:

$$Cost = \$100M + (3)(\$35M) = \$205M$$

The added cost of mitigating the electronic threat is:

$$Cost = \$80M + (3)(\$60M) = \$260M$$

And the cost of the ground segment is:

$$Cost = (3)(\$50M) = \$150M$$

The total cost is then:

$$Total\ cost = \$1.8B + \$205M + \$260M + \$150M = \textbf{\$2.4B}$$

**Option 2:** The cost of 12 identical satellites capable of delivering 30 Gbps is:

$$Cost = (12)(\$200M + \$90M) = \$3.5B$$

Adding the electronic threat mitigation, as in Option 1, increases the cost by:

$$Cost = \$80M + (12)(\$60M) = \$800M$$

Adding the cost of the ground segment:

$$\text{Cost} = (6)(\$65M) = \$390M$$

$$\text{Total cost} = \$3.5B + \$800M + \$390M = \$4.7B$$

This does not include any additional cost of commercial services if diversification is employed as well.

**Option 3:** Using the Option 1 architecture, the cost of the satellites and launch is identical: $1.8B. The added cost for reconstitution of all three satellites is an additional $1.8B. The cost of the electronic mitigation is higher due to the added cost to three additional replacement satellites.

$$\text{Total cost} = (2)(\$1.8B) + (\$80M + (6)(\$60M)) = \$4B$$

For this exercise, Option 1 provides the performance and resilience at the lowest cost. Though it is evident that the cost trade outcome is highly dependent on the actual economics of space, launch, and ground. For example, if the cost of a smaller satellite capable of delivering at least 2.5 Gbps of capacity could be acquired for only $100M instead of $200M, with dual launch capability, the Option 2 cost would drop to $2.5B, which begins to become cost competitive with Option 1.

The summary results of this example exercise are shown in the table in Figure 7.25. Using these notional cost values, Option 1 is clearly the best option in terms of providing the greatest cost efficiency versus the resilience achieved.

It is clear that the results are sensitive to the relative costs and amount of mitigation provided for each feature. These examples illustrate the importance of defining and analyzing the relationships between performance, cost, and resilience and ensuring that the chosen weighting factors represent true sensitivities that reflect the value to the end users of the system. While the mathematics might not be complex, the architecture, element, and mitigation options and the manner in which the threats interact with the space system could be, and that is the true value in creating accurate threat and system models for the purposes of optimizing the design.

| Architecture/Mitigation | Capacity (Gbps) | Resilience | Cost ($B) |
|---|---|---|---|
| Option 1: Protection | 30 | 0.91 | 2.4 |
| Option 2: Distribution | 30 | 0.73 | 4.7 |
| Option 3: Reconstitution (3-ball) | 30 | 0.77 | 4.0 |

**FIGURE 7.25**
Summary of trade results.

# 8

## The Future of Resilient Space System Design

Both the space system threat environment and base technologies continue to evolve and these changes will steer the future design trades for next-generation space systems. Today's design constraints are not necessarily the same as those of tomorrow. Solutions that today appear cost prohibitive or technically infeasible may become affordable and achievable. As a result, it is important to continuously monitor trends that materially affect the design process and trade outcomes to ensure that the solution space is refreshed and emerging solutions are not dismissed out of hand. As illustrated in Chapter 3 (Figure 3.6), a key to maintaining a resilient system is anticipating threats and continually modernizing space systems to ensure that they continue to be effectively mitigated.

The rapid evolution of the key technologies that define space systems has led to the deployment of very highly complex commercial and government space systems. The capacity of a typical commercial communications satellite in the 1980s was a small fraction of that of today's high-capacity and high-throughput satellites. Likewise, the advent of on-board digital processing, high-density high-speed semiconductors, phased array and multi-beam antennas, high-resolution sensors, and high-efficiency power amplifiers and solar cells has led to incredible advances in almost every type of satellite.

Meanwhile, digital computing technology, both hardware and software, has revolutionized the ground segment enabling the transmission and processing of enormous amounts of bandwidth and ever higher performance due to advanced digital signaling techniques. Even as RF spectrum has grown scarcer, new bandwidth-efficient technologies have enabled operators to squeeze ever more capacity through the same sliver of spectrum. Future demands for bandwidth may cause exploitation of frequency bands at higher frequencies as well as optical wavelengths.

Many of these advances are likely to continue and further expand the design choices that system designers have at their disposal to create more resilient systems. This is necessary because many of today's threats are anticipated to escalate with greater frequency, variety, and severity. The ability to remain resilient to these threats is crucial for maintaining relevancy in the future, as a system that can be easily disabled is unlikely to be built. How will future changes in the space environment, including technology advances and market pricing for capabilities, affect the design trades in the future?

To that end, this chapter discusses factors and trends that impact resilience design trades, including emerging technologies and threats and the resulting influence on the selection of threat mitigation options. There is clearly a cause and effect at work, and the better that the sensitivity to these unfolding developments can be understood, the more likely that these trends can be exploited in the future to continue to build more resilient space systems.

For any new tool to be useful to the system designer, it should enable an affordable and achievable increase in one or more resilience attributes without compromising significant performance. Each of the following trends are considered through this prism, linking the opportunities presented to one or more of these core attributes with the goal of obtaining a more resilient system.

## 8.1 The Cost of Satellite Capability on Orbit

A fundamental measure of any space system is the cost of acquiring a specified level of capability on orbit. This can apply to communications, navigation, imaging, and other missions. With the trend toward building more distributed architectures from a larger number of satellites, the desire is to place more capability on smaller satellites, or *smallsats*, at a lower cost. The cost is a combination of three key contributors: the cost of access to space (launch costs) for those satellites, the cost per satellite for a given capability level, and the cost of the ground segment, including mission planning and command and control elements. For satellite communications systems, the impact to the user terminal community is also a large component of the cost.

### 8.1.1 Launch Costs

Launch costs have driven space system economics since the birth of satellites, usually contributing significantly to the total system cost. Even in the twenty-first century, launch of a medium launch vehicle (MLV) to take a mid-sized satellite to geosynchronous orbit costs $90M to $200M, depending upon the launch vehicle configuration and provider, excluding launch insurance. For satellites costing $150M to $300M, this is a significant cost component. Even when smaller satellites can be launched in multiples per launch vehicle, the launch cost per satellite is still significant.

Since 2010 the launch market has expanded as more entrants have appeared. Space Exploration Technologies Corporation (SpaceX) introduced its Falcon 9 MLV for commercial launches at a price of approximately $56M in 2010, roughly half of the price of its nearest established competitor. Since then other companies have announced their intention to enter the market with similar lower cost offerings and this growing competition has led heritage providers to reduce their costs for some launches. The cost per kilogram of mass placed into a low Earth orbit (LEO) has decreased from approximately $25,000 in the 1980s to less than $3000 for the Falcon 9 in 2017 [13]. SpaceX's arrival has disrupted the commercial launch market, and the company is now launching U.S. government satellites as well. If other entrants are as successful, a permanent effect on market pricing is likely, with much lower cost per unit mass for satellite launches expected.

In addition, a wider range of options for launching smaller satellites have appeared. One example is the Evolved Expendable Launch Vehicle (EELV) Secondary Payload Adapter (ESPA) [14]. This payload adapter ring flies on a U.S. Delta IV, Atlas V, or Falcon 9, and was developed by the U.S. Air Force Research Laboratory (AFRL) to provide a more affordable means of launching smaller payloads, such as smallsats. To date, multiple missions have flown since the initial successful deployment in 2007 with the launch of the STP-1 satellite as the primary payload. While other smaller launch vehicles have been available, including the Minotaur, a derivative of the Minuteman and Peacekeeper missiles, the ESPA ring provides more flexibility for smaller payloads.

The impact of reduced launch costs can affect the design for resilience in multiple ways. First, reduced launch costs encourage the development of systems with more satellites, taking advantage of more distributed architectures. Budget that formerly was used to launch satellites can now be shifted to buy more satellites or invest in resilience features. Consider an operator with a $1B budget and current costs for a combination of satellite and launch vehicle total $250M. The operator can afford four satellites, assuming the size and capability of such a system meets its needs. Now suppose the launch costs decrease such that the per satellite cost is halved. Now the operator can provide a more distributed, resilient system of eight satellites with the same capability for the same cost. Figure 8.1 illustrates the resilience of a system to a threat causing a loss of two satellites versus the cost per satellite on orbit. If the per unit cost were to drop by a factor of 4, the resulting 16-satellite constellation could provide a resilience of 0.88 versus the original 0.5 that the four-satellite system provided. Given a fixed amount of funding, all else being equal, the cost reduction ratio can result in a similar ratio for creating a resilient distributed space layer.

Reduced launch costs also make the use of reconstitution more practical, in the form of responsive replacement of disabled or lost satellites. If the cost of having a launch vehicle and a replacement satellite on standby becomes more affordable, then a fast callup becomes a much more viable solution. As was presented in Chapter 5, one method of valuing reconstitution is in the time to reconstitute. So, having a system architecture that incorporates rapid reconstitution (relative to the operational timeline) also results in higher resilience.

If the loss of one satellite from a four-satellite constellation yields a resilience of R = 0.75, but the value of reconstitution is such that restoration of

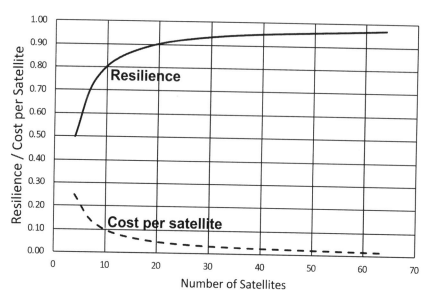

**FIGURE 8.1**
Resilience and cost per satellite versus number of satellites.

satellite operation within a short time period provides an $R_{RC}$ value of 0.5, then the new, adjusted resilience value is:

$$R = 0.75 + (1 - 0.75)(0.5) = 0.875$$

which is an improvement over system designs that do not include rapid reconstitution. On-orbit sparing could also provide another means for reconstitution, with the spare brought into service only when another satellite is lost.

If multiple smaller satellites can be launched on a single, responsive launch vehicle, then the effective reconstitution value may be even higher. All of this is based on an equivalent cost scenario, otherwise there is a cost differential that must also be considered in the overall trade. But clearly the trend of declining launch costs, if it continues, affects the options and trades for resilient system design.

## 8.1.2 Increasing the Capability Density and Affordability per Satellite

The amount of capability provided by individual satellites is key to any space segment trades in a resilient design. One key future trend to monitor is the amount of capability that can be placed on increasingly smaller (and cheaper) satellites. This capability can be communications capacity, imaging quality and quantity, navigation accuracy, simultaneous coverage areas, or any number of others.

Small satellites have been in the spotlight for several years now, with many commercial companies all vying to put up so-called *mega-constellations* of hundreds or thousands of smaller satellites, most in LEO or MEO orbits. The idea is that through high-rate production these smaller satellites benefit from high-volume economics to radically reduce the cost per satellite. While individually less capable than their larger siblings, these satellites nevertheless can provide sufficient capability and performance to meet the needs of a large, complex system which in aggregate fulfills the overall system requirements.

The success of this business case rests both upon the cost of access to space and the cost per capability per satellite, as well as the ability to achieve these targets by putting as much capability on as small and inexpensive a platform as possible. As the technology enables lower price points, though, the demand for capability increases, and the requirements often increase to meet it. The challenge is to deliver high enough value such that the system is not obsolete over its lifetime and continues to generate the revenue (for commercial systems) required to maintain financial solvency.

As with the launch cost equation, the more capability that can affordably be placed on smaller satellites enables system designers to potentially create a more distributed, resilient architecture which is robust at the system level. At some point the cost to disable a single satellite is considerably more than the cost of the satellite itself, and the incremental loss in system capability is so low as to render the threat's impact negligible. As with the cost of launch, smaller and cheaper satellites can result in more responsive reconstitution as well, providing greater resilience through that means.

The key is putting enough capability on a small enough satellite that it is still performing its mission or delivering its service. Figure 6.3 in Chapter 6 illustrated that there are practical limits that exist both in terms of satellite size and cost. Currently, most smallsats are limited in their ability to carry payloads that include large sensors or antennas. A smaller antenna, for example, limits the communications capacity of the satellite. Smaller sensors have reduced resolution and this performance might not meet mission needs. Smaller, less complex satellites may also lack the ability to adequately detect and avoid a threat due to a simpler, less expensive design, limiting their elemental resilience to certain threats.

The factors that most limit the amount of capability that can be placed on a satellite are the amount of power that their solar panels can generate (and energy their batteries can store) and the amount of heat that the thermal subsystem can radiate into space, as well as the size and mass available to host the payload. Medium power satellites supply 6 to 10 kW of payload power, while smallsats typically only provide 100 to 1000 W of available power. But the power efficiency of both solar cells and DC power subsystems continues to increase even as the efficiency of the payload that consumes it also is increasing due to more highly efficient RF and digital technologies. As a result, the amount of capability that a given sized smallsat can host continues to rise. Physics provides some practical limits, such as the aperture size of an antenna or telescope, or the transmit power for an RF amplifier. But small-sats offer promise in augmenting systems of larger satellites or as the building block for highly distributed mega-constellation systems.

Smaller satellites also tend to have much shorter on-orbit lifetimes, at least partly due to the lack of on-board redundant subsystems. Many satellites' mission life spans are limited by the amount of propellant that they contain. Once the propellant is expended, there is no way to maintain the position and orientation of the satellite and it becomes unusable. Obviously smaller satellites are more severely constrained by their ability to house propellant, even though they also use less to maintain station keeping on orbit. This also impacts resilience from the aspect of how often a satellite constellation is replenished. Shorter replenishment times can be leveraged to provide more rapid technology upgrades to the satellites to keep pace with evolving or changing threats, but also incur additional and more continuous replenishment expenses.

### 8.1.3  Cost of Increased Ground Complexity

As operators pursue larger and more complex satellite architectures, the cost of planning and managing them is also likely to grow. The commercial satellite industry has already demonstrated highly automated and efficient SOCs that can manage tens of satellites simultaneously with very little human intervention. But there are additional costs for systems that can accomplish this versus systems that must only manage a small number of satellites. As emerging systems exhibit more network-centric properties the complexity of the ground segment is likely to grow, too. The degree to which the costs increase is dependent upon the continued development of scalable, modular hardware and software to support space system management.

## 8.2  Space and Ground Segment Flexibility

For larger, more expensive satellites with longer life spans, the amount of on-board flexibility that can be incorporated into the bus and payload becomes important. This is also true of the ground segment, although it is easier to make periodic, incremental upgrades to the ground hardware and software than it is for a satellite on orbit.

As additional capabilities are added to satellites, including resilience features, complementary capabilities must often also be incorporated into the ground mission planning, command, and control hardware and software. More complex payloads require more complex configuration commanding and provide larger volumes of telemetry that must be processed. As a result, there is a proportional ground cost associated with advances in the complexity of satellite payloads and spacecraft. Designing ground systems that can be modularly implemented with standard interfaces reduces the cost and time required to make modifications as the space segment evolves.

Flexibility is difficult to explicitly define, but generally it is used to describe the ability to reconfigure, repurpose, and/or reprogram an item to change its behavior, capabilities, and/or performance. In the case of a satellite, this usually refers to the payload which contains the mission functionality. From a resilience standpoint, this is again an exercise in trying to stay ahead of an escalating or evolving threat once the satellite is on orbit. Consider a cyber threat, in which a day zero exploit to the cryptographic codes protecting the uplink commanding has been discovered. This vulnerability could provide an adversary with a convenient means of taking control of, or disabling, a satellite from a clandestine ground station. If the code is ubiquitous, multiple satellites could be at risk. If the operator discovers the exploit, then it would be highly desirable to be able to upload a new cryptographic algorithm or to patch the vulnerability. This is an example of flexibility being used to maintain the resilience of the satellite to a cyber threat.

Certain satellites include digital processors in what is known as a "regenerative payload," meaning that the received signals are demodulated and decoded on board the satellite and then, usually after switching and routing of the baseband (source) data, translated back into the analog domain and transmitted on the RF downlink. Flexibility for such a payload could include the ability to select among many different digital processing options to increase performance, anti-jam robustness, or enhance other capabilities.

These features could include the choice of forward error correction codes, interleaver depths, on-board encryption, and frequency hopping algorithms. While some of these features are already available, being able to update entire algorithms from the ground might be the next step in achieving greater resilience. Many digital processors for space are built using *application-specific integrated circuits* (ASICs), which are designed for very specific uses and are favored due to their high density and relatively low power. The use of *general purpose processors* (GPPs) executes software code in non-volatile memory that can be modified using commanded software uploads but does not possess the level of processing power available in ASICs and consumes more power. A third option, the digital *field programmable gate array* (FPGA) is a programmable integrated circuit that can mirror ASIC functionality, but with greater size and higher power. Some FPGAs can only be programmed once, limiting their flexibility, while others can be reprogrammed multiple times. In the future, some combination of all three products could be used to maximize flexibility and on-board resilience to multiple threats.

Even further in the future, entire so-called software-defined payloads (or satellites) could be the ultimate in on-board flexibility. In this case, received signals are immediately converted into digital data and all on-board signal processing could be performed in high-speed general purpose processors completely in software, which could be uploaded from the ground. Instead of selected signal processing algorithms, the entire satellite's payload behavior might be reprogrammed as needed. Both commercial and military satellites have already been moving in this direction, but the need for much greater processing power with higher power efficiency has limited the amount of software flexibility that can be implemented at this time, including full networking functionality for high volumes of data.

The ground segment has fewer limitations and indeed ground systems employ much greater levels of data processing in support of space systems. Mission planning and data processing sites increasingly are built using architectures similar to the cloud computing data centers pioneered by large Internet companies such as Google and Apple. Today's computing architectures are highly scalable and with advanced networking can be both flexible and resilient to a number of threats. The greatest threat to the ground today is the cyber threat, as many ground segments have large attack surfaces and are not "air gapped" from external networks. The protection comes often in the form of firewalls (both hardware and software) and other cyber resilience techniques.

Here flexibility for resilience involves rapid detection of intrusion and disruption for fast so-called "failovers" in which processing and operations functions are moved to another, unaffected part of the infrastructure which is isolated from the area under attack. Today's computing technologies, including virtualization and the use of hypervisors to create virtual machines and administer them individually, provide significantly increased cyber resilience. Conventional threats to ground facilities also need to be mitigated, particularly in areas in which security could be compromised. In this case, flexibility could include the ability to quickly move traffic from one gateway site to another with minimal impact to system users.

## 8.3  The Impact of Increased Congestion

Another trend that is unlikely to slow is the increasing congestion in space and on the ground. As more and more commercial companies and nations enter the space arena, the congestion is increasing. This congestion takes two forms: physical congestion in space in the form of satellites and orbital debris, and RF congestion resulting in unintentional interference to satellite communications links.

As recently as 2016, over 15,000 human-produced objects were being tracked in space. This number only includes objects large enough to be tracked; the actual number is much, much larger. Even without the introduction of mega-constellations, the launch rate for satellites continues to increase (Figure 8.2) [15], leading to an increase in launch debris in addition

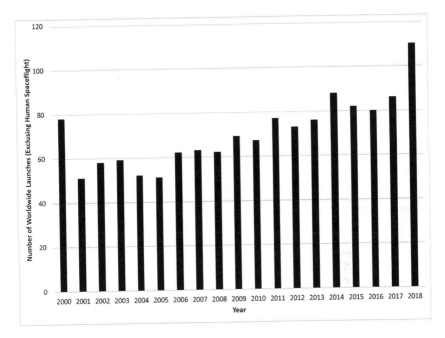

**FIGURE 8.2**
Worldwide launches by year (2000 to 2018).

to operational satellites. The abundance of objects in space increases the probability of collisions that can damage or disable operational space system assets. Even relatively small objects traveling at high velocities can cause significant damage to a spacecraft. These impacts represent an increasing threat to the entire space community, justifying increased interest in more detailed orbital analyses to better characterize the threat [16].

The abundance of kinetic threats in space could lead to the increased consideration of alternate orbits and the development of greater *space situational awareness* (SSA) to identify, detect, and track debris to increase the resilience to this threat in the future. Early warning is essential and is already employed to safeguard the International Space Station (ISS) from higher probability of intercept threats. The challenge is to develop indications and warnings (I & W) systems that can provide similar performance for smaller objects, perhaps using more sensors in space.

Physical threats are not the only threats that are multiplying. As more space systems are deployed, the radio frequency congestion also increases, leading to a greater incidence of unintentional interference of command and control and data links between space and ground. This electromagnetic smog can result in non-responsive systems and render communications channels useless. The number of large ground stations with powerful signals has already greatly increased along with hundreds of thousands of very small aperture terminals (VSATs) that have larger beamwidths and thus spread their transmit power over a wider swath of space.

All of these developments combine to contribute to a higher level of electromagnetic interference, increasing the effective noise level and making communications among space system elements more difficult if not impossible depending upon the circumstances. This trend might prompt a move toward designers using more robust communications techniques, including greater use of spread spectrum waveforms, new frequency bands, and perhaps eventually transitioning to such alternatives as optical (laser) communications.

## 8.4 Autonomy and Cognitive Systems

Many threat mitigation techniques are predicated on awareness and detection of the threat and a rapid response to either ensure system capability is not lost or to reestablish any lost capability as quickly as possible. Today much of the recognition and decision-making is performed on Earth by system operators with the support of systems with limited autonomy. These autonomous systems are often designed to look for very specific deviations from "baseline" or "normal" system behavior, helping warn operators when system performance exceeds pre-defined limits by sounding alerts and alarms. As space systems continue to increase in complexity and threats keep step, the level of system intelligence and level of autonomy granted must increase to keep pace, collapsing the response time to ensure an advantage.

Recovery is one of the most often used methods of achieving resilience. As such, the value of the recovery technique is not only based on the ability to recover capability as fully as possible, but also to do so as quickly as possible. Space systems are unique in that key elements of the system are unreachable at 22,000 miles above the Earth. An errant command could send a $250M satellite into a flat spin, losing the asset. Historically, operation of space systems has been a very labor-intensive endeavor. More recently the ground systems have become increasingly automated. This is predicated on the system being in normal operations for the vast majority of its operational lifetime. The operational tempo can change greatly in the event of an anomaly, when operators must carefully determine the source of the anomaly, its system impact, and the appropriate response required to regain nominal operation.

Simply put, persons in the loop for satellite systems operations slow response times due to the risk of making the situation worse. This can extend the outage time for system users while the root cause is found and corrective action is taken. The longer the outage, the less resilient the system, as has been shown previously.

One method of improving response times is to embed greater autonomy and cognitive functions in the system itself. A smart system can more efficiently self-monitor and quickly respond to suspicious or outright threatening conditions. Systems that can autonomously detect issues and perform self-healing operations will be able to do so considerably faster than human operators.

The concern with embedding higher cognitive functions, including some version of artificial intelligence, into space systems is that these systems generally set themselves apart from autonomous systems in that they learn and improve. But learning requires occasional failures and space is an unforgiving environment resulting in very conservative risk postures. Determining how to best train a cognitive engine and when to allow it to become operational is an important question with real consequences. However if smaller, less expensive satellites dominate in the future, then the impact of losing one may in some ways mitigate the risk of using advanced cognitive systems to operate them. A second issue is determining what level of policy authority these systems are given, and how much trust is imparted to faithfully execute operators' wishes, particularly in contested environments. These concerns must be addressed if future space systems are to fully embrace complex cognitive autonomous features.

## 8.5 Extension of the Terrestrial Network

A goal of many commercial and government SATCOM systems is to extend existing terrestrial networks, such as the Internet. Most existing satellite systems still provide "bent pipe" transponded relay of data placed on radio signals to move the data from one place to another. Adding digital processed payloads to satellites can turn them into actual network nodes, processing the baseband data and routing it on board to its destinations similar to conventional network routers. Many of the "new space" non-geostationary orbit (NGSO) systems are focused on providing wireless Internet access to areas of the world that are currently underserved.

As networking functions migrate to space, they are likely to carry with them some of the same vulnerabilities that currently exist on the ground. This opens up a new cyber threat to these systems, as the attack surface has been increased to include each space node. In order for the satellites to provide the network functions, they need to be compatible with terrestrial standards, such as the Internet Protocol (IP) and others as well. As a result, exploits that affect ground networks are also likely to be applicable to space nodes if the attacker can find a path to access them.

These new cyber threats require system designers to be diligent in their software architecture design to ensure that proper cyber security used primarily on the ground is migrated to the satellite payloads as well. In addition, the satellite network nodes could provide new paths throughout the entire system, space and ground, which formerly did not exist and requiring the security approach to the entire system to be reviewed and modified based on this new construct.

Finally, the network functionality built into the satellite payload might need to be reprogrammable from the ground to ensure that it can be patched, as with ground software, in the event that a vulnerability (such as a zero-day exploit) is discovered. The drawback of such an approach is that if an adversary does penetrate the satellite, it could be able to upload its own software to further its own objectives.

## 8.6 On-orbit Servicing

The concept of on-orbit servicing has been the subject of both government and commercial research and development for several decades. On-orbit servicing enables a satellite to be serviced, including some amount of reconfiguration, in space. In most scenarios a servicing spacecraft autonomously docks with the satellite to be serviced and performs a set of pre-planned tasks. These tasks can include refueling, hardware upgrades, or any number of other functions.

For an advanced case, significant bus and/or payload upgrades could be made to increase not only the performance and capability of a satellite, but also its resilience. Threat mitigation features could be continually upgraded to ensure parity with an escalating threat environment. Properties of the satellite could be changed to render it less vulnerable to specific threats.

This model often requires changes to current spacecraft designs to provide features such as docking ports to enable the service spacecraft to perform their mission. Development of both the service vehicle(s) and the serviced satellites must thus be developed concurrently to ensure success. Some more recent approaches do not require such modifications, though.

On-orbit servicing is not yet commercially available, but agencies such as the U.S. Defense Advanced Research Projects Agency (DARPA) have shown that the concept is viable and limited on-orbit demonstrations with prototype spacecraft have been used to validate the concept. In the not-so-distant future these services might provide yet another option for system designers to extend the lifetime and usability of a space system through modification of on-orbit assets, with even greater flexibility than is currently available. The availability of these options might further tilt the design trade in the future towards incremental upgrades in space.

On-orbit servicing can increase resilience through increased avoidance, robustness, and recovery depending on the mitigation features that are being upgraded or installed. Adding a new propulsion system could improve the maneuverability of the spacecraft enabling it to avoid a kinetic threat.

Addition of sensors could provide early warning features so that a spacecraft can avoid space debris. Much of the fundamental technology has already been demonstrated but has not yet been integrated into new spacecraft.

Commercial business cases have not yet been proven that justify such investments and satellite manufacturers must also make changes to their product lines to enable such operations. Governments also must find insertion points into continuing missions that provide value for the investments that are required to adjust to this new approach to satellite longevity. And other satellite subsystems must be examined to ensure that they do not become the limiting factor in satellite lifetime even though sufficient fuel is maintained.

Nevertheless, it is likely that as certain applications of on-orbit servicing eventually prove to be cost effective, they can also bring resilience benefits to both commercial and government space systems. Extending the useful life of satellites via on-orbit servicing is an emerging area that bears watching as a future industry that can be leveraged by space system designers to provide additional resilience if the price is right.

# References

1. Pawlikowski, E. et al., Space: Disruptive Challenges, New Opportunities, and New Strategies, *Strategic Studies Quarterly*, Spring 2012.
2. Continuity of Operations (COOP) is a U.S. federal government initiative, required by U.S. Presidential Policy Directive 40 (PPD-40), to ensure that agencies are able to continue performance of essential functions under a broad range of circumstances.
3. U.S. Department of Commerce National Telecommunications and Information Administration, "United States Frequency Allocations: The Radio Spectrum," https://www.ntia.doc.gov/files/ntia/publications/january_2016_spectrum_wall_chart.pdf.
4. Washington Post, "Chinese hack U.S. weather systems, satellite network," November 12, 2014.
5. U.S. Department of Defense Fact Sheet, Resilience of Space Capabilities, 2011.
6. Office of the Assistant Secretary of Defense for Homeland Defense & Global Security, Space Domain Mission Assurance: A Resilience Taxonomy, A White Paper, September 2015.
7. Fetter, S., "Protecting Our Military Space Systems," in Edmund S. Muskie (ed.), *The U.S. in Space: Issues and Policy Choices for a New Era*, (Washington, DC: Center for National Policy Press, 1988), pp. 1–25.
8. Poisel, R., *Modern Communications Jamming Principles and Techniques*, Artech House, Boston, MA, 2011.
9. Morgan, F. E., *Deterrence and First-Strike Stability in Space: A Preliminary Assessment*, RAND Corporation, 2010.
10. Linkov, I. & Kott, A., *Cyber Resilience of Systems and Networks*, Springer Publishing Company, New York City, NY, 2018.
11. Rohani, H. & Roosta. A. K., *Calculating Total System*, Information Services Organization Amsterdam, [Online]. Available: https://www.os3.nl/_media/2013-2014/courses/rp1/p17_report.pdf, 2014.
12. Messer, Jr., G. H., *The Allocation of Availability Parameters – Repair Times and Failure Rates*, Texas A&M University, 1970.
13. Jones, H. W., "The Recent Large Reduction in Space Launch Cost," 48th International Conference on Environmental Systems, 2018.
14. Perry, B. "ESPA: An Inexpensive Ride to Space for Secondary Payloads." July 2012 Edition of *Milsat Magazine*. Retrieved September 26, 2012.
15. Space Launch Report website: http://spacelaunchreport.com/logyear.html.
16. Kelso, T., Analysis of the 2007 Chinese ASAT Test and the Impact of Its Debris on the Space Environment, *Proceedings of the Advanced Maui Optical and Space Surveillance Technologies Conference*, held in Wailea, Maui, Hawaii, September 12–15, 2007.

# Index

Printed in the United States
by Baker & Taylor Publisher Services